# Earth & Space

**Stephanie L. Redmond**

Illustrated by David W. Taylor, Jr.
and Leslie L. Northcutt

Dover, DE

*Christian Kids Explore Earth & Space*
by Stephanie L. Redmond
A part of the Christian Kids Explore series

Published by Bright Ideas Press
P.O. Box 333, Cheswold, DE 19936
www.BrightIdeasPress.com



Cover illustration and Coloring Pages by David W. Taylor, Jr.
Interior illustrations by Leslie L. Northcutt
Maps by Tyler H. Hogan

*Cover and interior design by Pneuma Books, LLC*
Visit www.pneumabooks.com for more information.

Library of Congress Cataloging-in-Publication Data

Redmond, Stephanie L.
Christian kids explore earth science / by Stephanie L. Redmond.
p. cm.
ISBN-10: 1-892427-19-2 (pbk. : alk. paper)
ISBN-13: 978-1-892427-19-9 (pbk. : alk. paper)
1. Earth sciences 2. Christian education—Activity programs. I. Title.

QE29.R385 2006
550—dc22

2006024064

11 10 09 08 07 06 6 5 4 3 2 1

*This book is dedicated to my pastor/teacher and friend, Dr. Doug Munton, First Baptist Church, O'Fallon, Illinois, and to my former pastors and forever friends, Ron David and Rick Hensley, Faith Community Church, Camden, Delaware. I thank God for the influence you've each had in my life, for your faithfully and rightly dividing (2 Tim. 2:15) the Word of God.*

# TABLE OF CONTENTS

## Appendices

# Acknowledgments

That God would even allow me the opportunity to teach anyone is amazing. Thank you to my Savior and LORD, Jesus Christ.

Linda Safford and Peggy Payne ~ You ladies hold me to the strictest adherence to God's Word. Thank you for the many years of encouragement, counseling, and true-blue friendship. Your Godly examples challenge me daily!

The Redmond women ~ Pam Metz, Susan Redmond, Allen Redmond, Alice Redmond. The teachers you all are and have been set a high standard that I will never attain. You inspire me greatly. Furthermore, though I hold no teaching degree, you acknowledge me as a fellow teacher. Thank you for all your encouragement through my homeschooling years.

Gwen and Joe Redmond ~ A very special thank you for all the encouragement, love, and support you have given Andy and me throughout the 23 years of our marriage. I love you!

Leslie L. Northcutt ~ My sister. I am so thankful that you joined this project. You are so smart and talented. Thank you for contributing the technical art for this book. I know you could easily have written this book yourself!

David W. Taylor ~ My brother. Once again, your talent amazes me. It will forever be a mystery to me...how you can take the lame descriptions and goofy sketches I give you and design the beautiful pages that you do! Thank you for letting God use your great gift on my humble project.

Melissa Craig and Beth Barr ~ My special thanks to both of you for compiling an excellent book and resource list. I truly appreciate your generous spirit and helping hands in taking on this challenge!

Andy ~ You have been so patient and so supportive during this effort. Thank you for the steak dinners, the laundry, and everything else! Your time donations enable me to do the things I believe God asks me to do. I am forever grateful.

Mike, Taylor, and Rachel ~ I love you so very much. Thank you for your Godly lives, obedience, and respect. I appreciate the way you each contribute to my projects. Your love and encouragement mean the most of all.

Maggie Hogan ~ my friend. Thank you for believing I could do this at all. Thank you for your patience. Thank you for helping me learn the joys of homeschooling and especially homeschooling through high school. So far, so good!

Liz and Bob Ridlon ~ I'm so glad we had the opportunity to live near each other again! Thank you for helping me keep my facts straight. Thank you for Sunday School and Bible Study and the great conversations we had, based on God's Word.

LTC (ret.) Frank M. Worley, USAF ~ Thanks to my "Pop" and favorite member of the American Meteorological Society for reading the weather and atmosphere units, helping to catch my mistakes and check my information. I will always remember your teaching us that the weather is not /supposed/ to do anything, but may be /forecast/ to do a specific thing. God is in control of the weather! I love you — You too, Neenie!

# A NOTE FROM THE AUTHOR

I am like you. I am an average homeschool mom, working hard to ensure a sound education for my children and still maintain a respectable home. I have three precious children: Mike (19), Taylor (16), and Rachel (14). My husband, Andy, and I have been married for 23 wonderful years.

My main goal is to glorify God in all that I do. Part of fulfilling that goal is teaching my children to love the Lord their God with all their heart, soul, and might. If I am going to do that, I believe it is essential to teach them from a Christian worldview, using books that honor God.

In our years of homeschooling, I have used several types of study. Some I liked, others I didn't. However, a few years ago, I was introduced to classical education. Now, I'll admit we're not perfect examples of this, but I found it to be a wonderful way to set up our school. We follow the trivium, recognizing the grammar, logic, and rhetoric phases of our children's learning, and we teach history chronologically. We use Latin in our studies and read, read, read! However, for our elementary students, we've had a hard time finding just the right science book. I felt forced, at times, to use books that contained teachings contrary to my beliefs—for instance, books that taught evolution as fact. While I was able to use this as an opportunity to teach my children God's truth, as presented in Scripture, I found that I longed for a homeschool book that would teach the same.

So it was that, through much prayer and encouragement from friends, I found myself writing my first homeschool science book, *Christian Kids Explore Biology*. You must know that

I am not a scientist. My credentials are that I homeschooled myself! I had the manuscript for the book reviewed and edited for science content by qualified people. During this process, I personally learned the joy of studying science. My hope was, and still is, that I will be able to impart that excitement to you and your children. Most of all, I pray that your family is encouraged and God is glorified through the words of the text that seek to exalt God as Creator.

Now I am very excited to bring you another book in the Christian Kids Explore science series, *Christian Kids Explore Earth and Space.* This is an earth science course and again, because I am not a scientist, I had the manuscript reviewed and edited for science content by qualified people. Hopefully, through this book, you can show your child the miracle of God's awesome Creation. Certainly, it won't be any great writing on my part that will do that. Rather, it will be through giving God the credit for the detailed structure and function of this planet. We will spend most of the school year, about 24 weeks, studying our Earth, leaving about 6 weeks to cover outer space.

Because I am much like you, I understand the pressures that homeschooling mothers and fathers face. Time is critical and there is never enough of it. Therefore, this book has been designed taking that into consideration. First and foremost, *Christian Kids Explore Earth and Space* **is written for multiple ages and grades.** While it is geared for 3rd to 6th graders, there are many ideas for younger ones, as well as for those who want to do more. The lessons are complete and concise, but there is room to bring in books from outside sources if you choose. Little advance preparation is needed. A list of the materials you will need is at the beginning of each unit so you can gather once per unit if you choose, instead of every week. Vocabulary lists are included, along with hands-on activities that reinforce learning. There is a gorgeous coloring page with each unit as well as a unit review. Also, there is a very helpful book and resource list in the appendix section. This book seeks to offer you everything you need for a fruitful year of elementary earth science, plus a little more.

~ Stephanie L. Redmond

*"Be exalted, O God, above the heavens;*
*Let your glory be above all the earth."*
*(Psalm 57:11)*

# HOW TO USE THIS BOOK

This book contains 24 lessons and six unit reviews. Each lesson and each unit review is designed to be completed in one week. If you teach science twice weekly, you'll need to allow for about 60 to 90 minutes each day. Of course, this will depend on the student and the number of outside resources used.

Each lesson consists of a **Teaching Time** and a **Hands-On Time**. I recommend doing each on a separate day.

## Teaching Time

- As each new lesson is begun, the text is read. You may read it to your students or they may read it to themselves. In the case of very young students, you might read it on your own and then discuss the information at their level. They may enjoy completing a Coloring Page while listening to you.

- After this lesson is read, students should complete a "Daily Reading Sheet."

To make the study complete, you will need to do a little more:

- First, review recent lessons, particularly as they apply to your newest lesson.

- Second, if you make flashcards as you go (with vocabulary words, lesson facts, Scripture verses, etc.), you will want to review those.

- Third, have your students list the vocabulary words (any in bold lettering in the lesson, plus any they listed on their Daily Reading Sheet) and define them in their science notebook.

- Last, you'll want to allow time for outside reading and picture perusing and researching topics of interest. I recommend having your students complete additional Daily Reading Sheets for their supplemental reading, even if they use just a few pages from a particular book. (It's quite acceptable to pick and choose pages and chapters to read rather than an entire book!) All completed forms and written work should be kept in their science notebook.

## Hands-On Time

Most children love hands-on learning, and it helps keep science exciting for your children (and you!). Although Hands-On Time can be time consuming, try to also make time for a little review as you are working. The "Checking It Out" science experiment form will often be utilized on these days and should be completed and filed in the student's science notebook.

It can be tempting to eliminate these activities to save time; however, I strongly advise otherwise. Science can be so exciting, but it can also be dull. It all depends on how it is taught. Elementary science is about discovery and taking joy in the journey. Have fun with it!

## Timeline Activity

Included at the beginning of each unit are timeline entries and dates. These can be used simply to add to your students' knowledge of events. However, I believe your students (and maybe yourself) will be better served by actually completing a timeline, adding information each unit. This activity will help them see major events, inventions, and scientists associated with the subject matter of each unit. The information listed will also give you and your students ideas for additional research. Wall-sized timelines are available from www.brightideaspress.com. We have utilized timeline activities throughout our homeschool studies and find them to be incredibly useful.

## Coloring Pages

There is one Coloring Page per unit and all of these are provided again in the appendix section for your convenience. These may be photocopied. Children of all ages will enjoy these beautiful drawings. Some will even benefit from keeping their hands busy with markers or pencils while having lessons read aloud to them.

## Unit Wrap-Up and Show What You Know!

The last event for each unit is a unit review and quiz. The "wrap-up" review gives students the opportunity to write a composition summarizing each lesson with the aid of the "Write About It!" worksheet and form (found in the appendix section). This form will help them think through and organize the unit material before they begin writing—all of which will reinforce the main points they studied in the unit. They are also encouraged to create a colorful and fun folderbook for each unit. (Folderbook instructions are simple and are included in the appendix section.)

The wrap-up quiz—"Show What You Know!"—can be used as a test or merely as a unit review; it's your choice. I've made the scoring fun, using thousands of points instead of one hundred. These, too, can be copied for each student and filed in his or her science notebook.

## Reproducibles

There are a number of forms in Appendix A that are available for reproducing, according to your needs. The course is designed to be easily used with several children of differing ages at the same time. If you do not have a home copier, make a trip to your favorite copy shop and reproduce several Daily Reading Sheets and Checking It Out forms (along with the Coloring Page, the Write About It! worksheet and form, and the Show What You Know! quiz for each unit). The first two forms mentioned are used frequently, so plan ahead. Is copying difficult for you? Make your own similar forms on the computer or simply use notebook paper. It's the content—not the form—that counts!

Also in Appendix A are labeled and unlabeled versions of the labeled technical art that is in the lessons. When you're at the copy shop, reproduce as many of the unlabeled versions as you'll need for each unit; then have your students test themselves by filling in the labels without looking in the book. When done, they can easily check their labels against the book. This provides another activity to reinforce material they have studied throughout the units.

# What a Daily Lesson Could Look Like

## Tuesdays

- **Memory Work**—Review flashcards and vocabulary. *5 minutes*
- **Discuss** last lesson. *5 minutes*
- **Teaching Time**—Read or have student read new lesson; ask comprehension questions as you go. *10 minutes*
- **Discuss** new information. *2–3 minutes*
- **Daily Reading Sheet**—Have student complete a Daily Reading Sheet. *10 minutes*
- **Vocabulary**—Have your student fill out (or assist him or her in filling out) a vocabulary sheet, or make flashcards, if you prefer, of the key words in the lesson. *10 minutes*
- **Books**—Outside reading time. This is where students have the opportunity to peruse other sources, perhaps from the library. *30 minutes or more, as necessary*

## Thursdays

- **Memory Work**
- **Hands-On Time**—Complete a relevant experiment or activity. Student can also use part of this time for working on folderbooks and discovery (research) activities.

**Remember:** Younger children do not need as much detail. Give them the facts and HAVE FUN! We are trying to include enough "work" for the older kids, but enough "fun" for the younger ones. If you have only young elementary (K–3) students, then even once a week is enough for science. If they are doing memory work, though, bring out those flashcards two or three times per week. We like to go to the library and get lots of books on the subject at hand.

## Instructions for Your Science Notebook

This section is addressed to your students; however, you may need to help them decide the best way to organize their science notebooks.

This year you will need to maintain a science notebook. The purpose of this notebook is to help you organize all your documents from your studies. An important part of good science is good record keeping. It is the only way to accurately track your findings.

**I recommend a three-ring, loose-leaf notebook, about 1½ inches thick, with pockets on the inside of the covers. For tabs, I recommend tabs with labels. You have two options in this area:**

- **Option 1**—Unit by Unit: For this method you will need six tabs labeled "Unit One," "Unit Two," and so on, through Unit Six. In each section you will file your Daily Reading Sheets, Checking It Out forms, Write About It! composition, and any other written work.

- **Option 2**—Type of Work: For this method you will need at least eight tabs, possibly more, and you will file your work chronologically, that is, in order by date. Your tabs should be labeled:
    - Daily Reading Sheets
    - Vocabulary
    - Checking It Out
    - Write About It!
    - Coloring Pages
    - Field Trips
    - Diagrams (You might sketch some from your readings.)
    - Photos (I highly recommend taking photos throughout the year of your Hands-On activities, field trips, and experiments.)

Your science notebook will provide an excellent record of your studies in earth science!

Christian Kids Explore

# Earth & Space

Rain
Change
Fair
Very Dry

# Unit One
# Getting Started

## Unit Timeline

250 B.C. Eratosthenes estimates the size of Earth.

1970 Earth Day—April 22—is established. Earth Day is intended to recognize the impact of humans on this planet.

## Unit One Vocabulary

- axis
- circumference
- core
- crust
- inner core
- lithosphere
- lower mantle
- outer core
- revolve
- rotate
- upper mantle

## Materials Needed for This Unit

- Science notebook
- Photocopy of the world map in Appendix A
- Photocopy of "Earth's Structure Pattern" in Appendix A
- Photocopy of the unlabeled "Structure of Earth" diagram in Appendix A
- Globe
- Colored markers or pencils
- Scissors
- Double-sided tape
- Modeling clay or play dough in five different colors (See the recipe for play dough in Appendix E.)
- Dental floss
- Camera
- Materials for Unit Wrap-Up (See page 190.)

Change
Fair
Very Dry
KA
JET SET

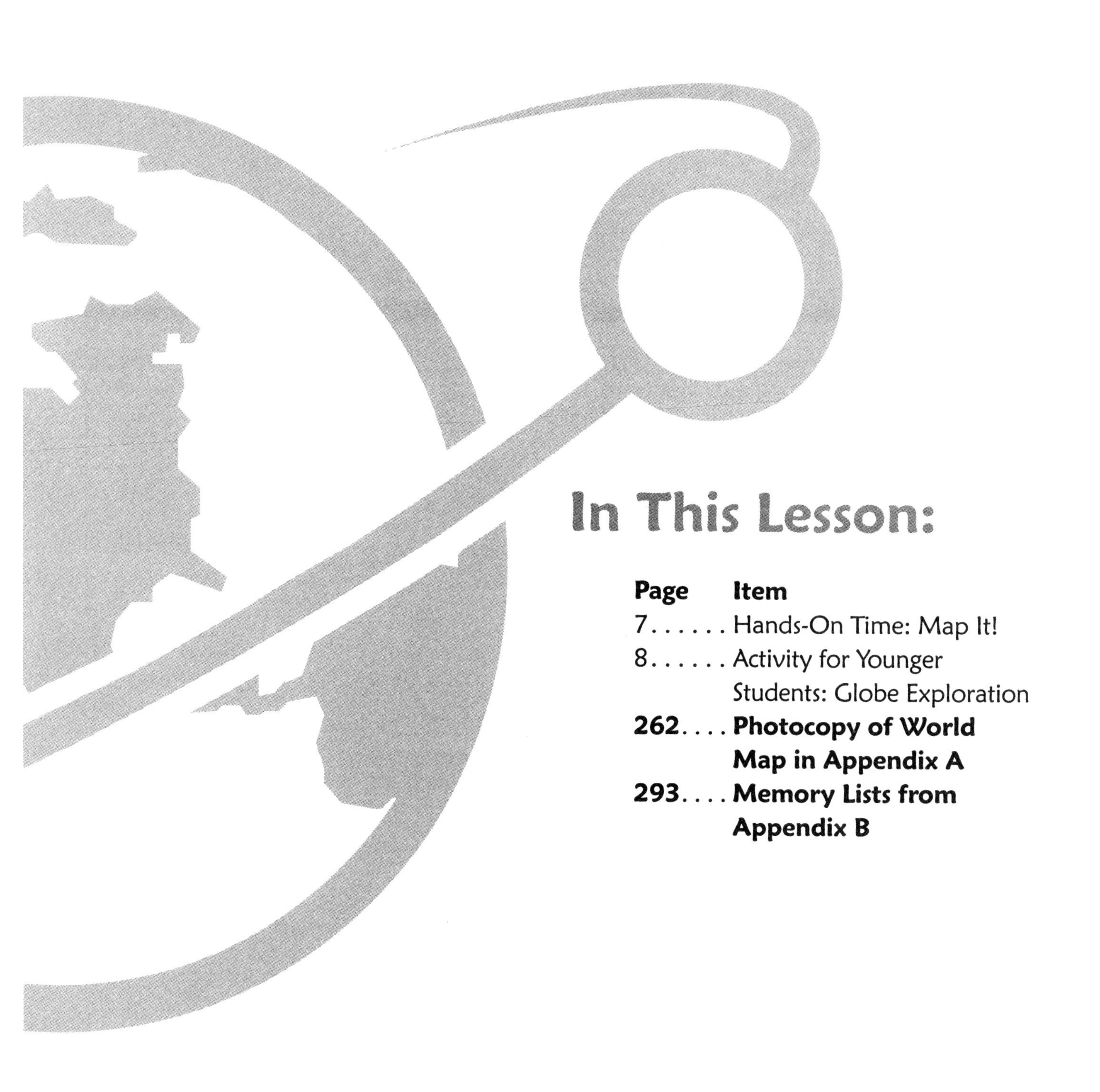

# In This Lesson:

# Lesson 1

# EARTH'S CREATION

## Teaching Time:
## In the Beginning...

God created the heavens and the earth. Did you notice that the Bible mentions "the earth," specifically? It does not say, "God created Mercury, Mars, Venus, Jupiter, Saturn, Uranus, Neptune, Pluto, and Earth. God created all the planets, but the Bible says, "the heavens and the earth." Wow! God must think that Earth is pretty special to single it out in that way. One of my favorite Bible teachers suggests that people are the reason Earth is so special to God. After all, He made us and loves us.

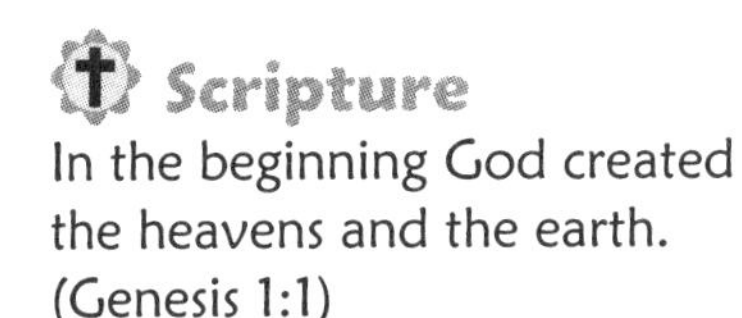

**Scripture**
In the beginning God created the heavens and the earth. (Genesis 1:1)

In my opinion, that gives us two very good reasons to study Earth. First of all, God created it and thought it was important enough to tell about in His written Word, the Bible. Second, we live here. We should be interested in learning about the place where we live. I don't know about you, but even though I like learning about other planets, I like learning about Earth the most. After all, it directly affects me.

Let's go back to Genesis 1 and see what else we can learn about the creation of Earth. Genesis 1:2 tells us four things

**Additional Notes**

about Earth when it was first created by God. "The earth was without form, and void; and darkness was on the face of the deep. And the Spirit of God was hovering over the face of the waters." Did you see the four things?

1. The earth was without form.
2. The earth was void.
3. Darkness was on the face of the deep.
4. The Spirit of God was hovering over the face of the waters.

The Bible also tells us about the days of creation. In your Bible read Genesis 1:3–31. In your science notebook, list what God created on each day. Next, look at Genesis 2:1, 2. What happened on the seventh day? Yes, God rested.

As we explore Earth and outer space this year, we will build on the things we have just studied. God is the source of all. According to the Bible, there is nothing accidental about Earth. It is God's masterpiece.

Before we end this lesson, let's look at one more thing. Genesis 1:1 says, "In the beginning . . ." This is not the only place in Scripture where we learn about creation. We can also look in John 1. John 1 also tells us about the beginning. Let's read it together.

> In the beginning was the Word, and the Word was with God, and the Word was God. He was in the beginning with God. All things were made through Him, and without Him nothing was made that was made. In Him was life, and the life was the light of men. And the light shines in the darkness, and the darkness did not comprehend it. (John 1:1–5)

"The Word" is Jesus Christ. When God created the world, He was there; His Spirit was there "hovering over the face of the waters"; and His Son, Jesus, was there. They are all part of one God, and all were present at the creation of Earth.

We will learn so much about the Earth that God created, as well as about the planets in space. Let's have fun and *explore*!

# Hands-On Time: Map It!

## Featured Activity

**Objective:**
To become familiar with the components of Earth.

**Materials**
- Globe
- Photocopy of the world map in Appendix A
- Memorization lists from Appendix B

**Instructions**

1. Locate the seven continents on the globe and list them in your science notebook.
2. On the world map, label the continents.
3. Locate the oceans on the globe and list them in your science notebook.
4. Label the oceans on your map.
5. Locate the equator on your globe.
6. Label the equator on your map.
7. Locate as many seas as you can and list them in your science notebook.

**Scripture**
To see for yourself that "the Word" in John 1 is Jesus, continue reading through verse 17 and see who John is talking about.

**Discovery Zone**
Check out the "Recipe for Planet Earth" at www.factmonster.com. Select "Science," then "Environment."

Additional Notes

8. Label the seas on your map.

9. Using the lists in Appendix B, make flash cards for memory work. Daily practice with these will help you memorize them more quickly.

## Activity for Younger Students: Globe Exploration

1. Use a globe to point out the continents and oceans.
2. Count how many continents there are.
3. Count the oceans.
4. Find the equator.

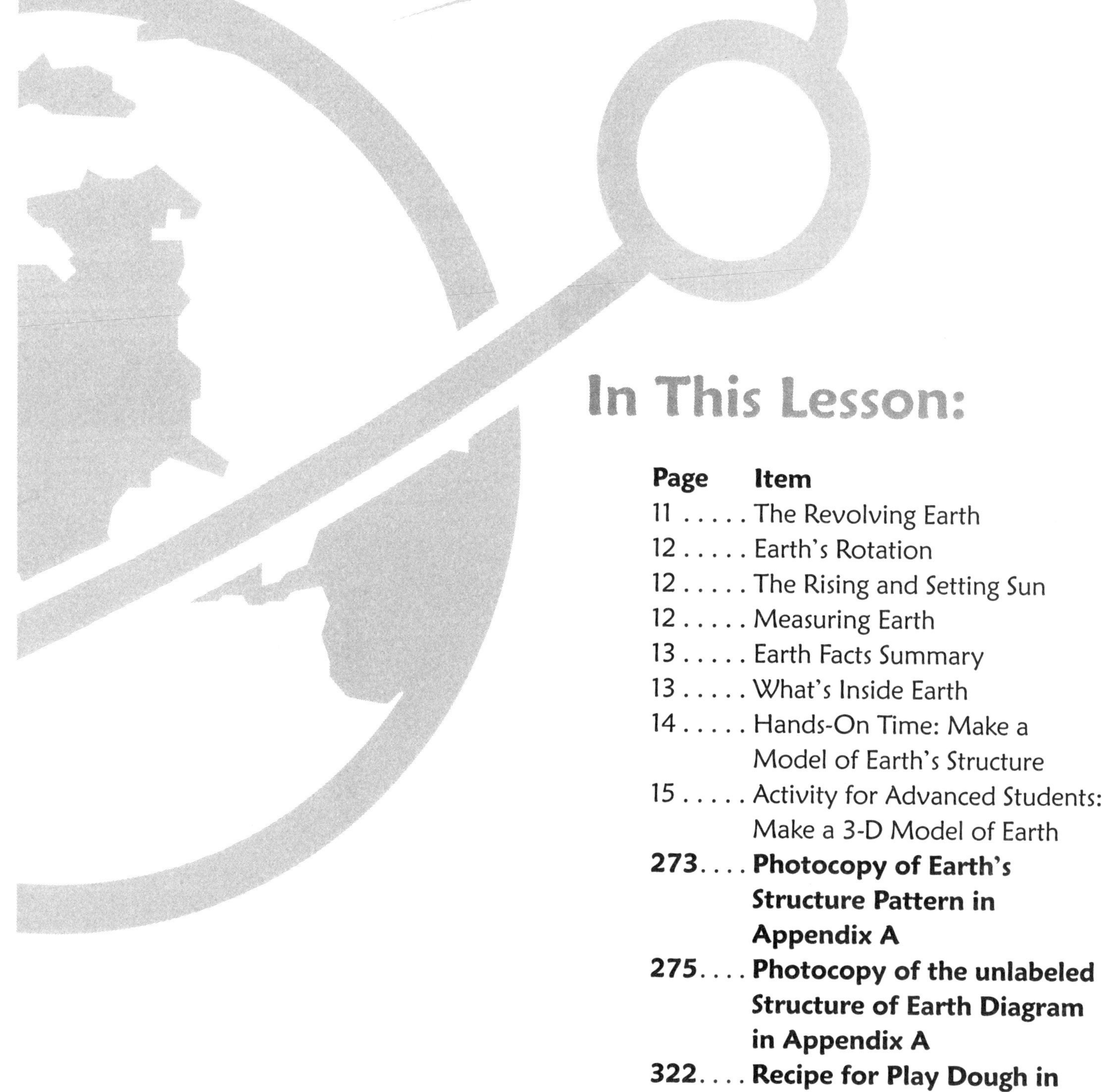

# In This Lesson:

Lesson 2

# FACTS ABOUT EARTH AND ITS STRUCTURE

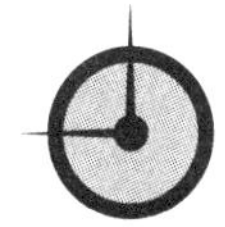

## Teaching Time:
## Just the Facts, Ma'am

In the last lesson, we looked at the creation of Earth and God's role as Creator. Today, we're going to study many facts about Earth. When I study these facts, I am more certain than ever that God created our whole universe with design and purpose.

### The Revolving Earth

Our Earth is one of nine planets that **revolve**, or move around, the Sun in a pattern. It is 93 million miles from Earth to the Sun. This distance is the exact distance necessary for water to exist in liquid form without boiling. If Earth were any closer to the Sun, life could not exist here because it would be entirely too hot. If Earth were any farther away from the Sun, it would be far too cold for life to exist. In other words, Earth is in exactly the right place. It is the only planet in our solar system proven capable of supporting life.

Scripture

"'Where were you when I [God] laid the foundations of the earth? Tell Me, if you have understanding. Who determined its measurements?'" (Job 38:4–5)

Name It!

**revolve**
To move around another object.

**Activity**

Make a play-dough ball to represent Earth. Insert a toothpick through the center of the ball so that it sticks out each end. This represents the axis. Tilt your whole Earth slightly and spin it around the axis. This represents rotation.

**Activity**

**"Revolving" Demo**

1. Place a chair in the middle of the room. The chair represents the Sun. You represent Earth.

2. Walk around the chair in a big circle. You are *revolving* around the chair.

3. Face the chair. The front of your body is having daytime.

4. Slowly spin in place. You are *rotating*. As you spin, notice where your body is facing. Stop with your back toward the chair. Is it still daytime for the front of your body? Why or why not?

5. Finally, making sure nothing breakable is in your path, combine walking around the chair (revolving) with your spinning (rotating) motion. Earth rotates on its axis as it revolves around the Sun.

## Earth's Rotation

In addition to revolving around the Sun, Earth **rotates** on its **axis**. Do you know what the axis is? Do you know what rotation is? Rotation is a spinning motion. The axis is an imaginary rod through the center of Earth around which Earth rotates. Did you get that? If not, do the activity listed in the margin. It will help you to understand rotation and axis.

Earth rotates at a speed of more than 1,000 miles per hour at the equator, taking 24 hours to complete one full rotation. This rotation is perfect. A different speed could cause us to be exposed either too long to the Sun, or not long enough. This would not be a good thing.

## The Rising and Setting Sun

You might be making a discovery right about now. You might be realizing that Earth's rotation causes our nights and our days. When Earth rotates, different parts of Earth face the Sun. Up until now, you may have thought that the Sun moves across the sky. You're not alone. Many children believe this because it actually looks that way. We may say "the Sun rises" and "the Sun sets," but in truth, Earth is the one moving. Earth rotates in an eastward direction, causing the Sun to appear to "rise" in the east and "set" in the west. We will explore all of this and more in our hands-on activity later in this lesson. For now, it's enough to work on learning the facts.

## Measuring Earth

Let's add a few more facts to our lesson. Earth's **circumference** (sir **come** fer unce) is its measurement all the way around the equator. The circumference of Earth is 24,870 miles. This makes it the fifth largest planet in the solar system.

These facts are all part of what makes Earth so special. As we revolve around the Sun and rotate on the axis, let's give glory to God for His wonderful creation.

## Earth Facts Summary

- Earth revolves around the Sun.
- Earth is 93 million miles from the Sun.
- Earth rotates on its axis at a speed of 500 miles per hour.
- It takes 24 hours for Earth to complete one full rotation.
- The circumference of Earth is 24,870 miles.
- Earth is the fifth largest planet in the solar system.

Now that you have learned some basic facts about Earth, it's time to go deeper—all the way to the center of our Earth!

## What's Inside Earth?

You learned that the circumference is the measure around Earth and that it is 24,870 miles. That is the same as traveling from New York to California and back again five times! Look on a map of the United States and locate New York and California. If you could drill a hole all the way to Earth's center (and you can't), you would find it to be 3,950 miles down. That's about one and a half times the distance from New York to California. Obviously, the distance around Earth is much more than the distance to the center of Earth, but both distances are quite far. Let's find out what's inside Earth.

The very center of Earth is called the **inner core**. The inner core is made up of solid iron. Around the inner core is the **outer core**. The outer core consists of liquid iron and nickel. We refer to both of these sections together as simply the core. The temperature at the core of Earth is 5,432 degrees Fahrenheit.

The section of Earth surrounding the core is the **lower mantle**. The lower mantle is made up of solid rock.

After the core and the lower mantle, there is the **upper mantle**, which is solid rock like the lower mantle. Around all of this, on the very outside of Earth, is the **crust**. The crust is about six miles thick. Together, the crust and the upper mantle form the **lithosphere**.

The cutaway "Structure of Earth" diagram in this lesson will help you better understand and visualize these layers in our Earth.

### Name It!

**rotate**
To spin on an axis, either real or imaginary.

**circumference**
Measurement around an object.

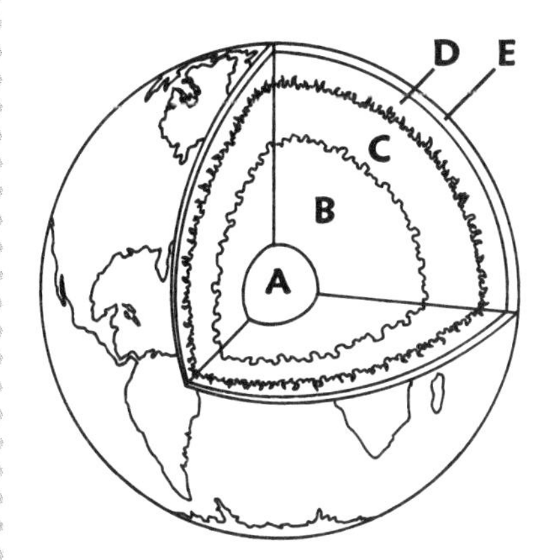


### The Structure of Earth

**A. Inner core**
The very center of Earth; made up of solid iron.

**B. Outer core**
The area of Earth that surrounds the inner core; made up of liquid iron and nickel.

**C. Lower mantle**
The area inside Earth that is solid rock and surrounds the core.

**D. Upper mantle**
The area inside Earth that surrounds the lower mantle and is made of solid rock.

**E. Crust**
Outermost layer of Earth; about six miles thick.

**axis**
Imaginary rod through the center of Earth around which Earth rotates.

**core**
The area inside Earth made up of the inner and outer cores combined.

**lithosphere**
The part of Earth made up of the upper mantle and the crust.

## Hands-On Time:
# Make a Model of Earth's Structure

## Featured Activity

### Objective
To learn the order of the layers in Earth's structure.

### Materials
- Colored markers or pencils
- Photocopy of "Earth's Structure Pattern" in Appendix A
- Scissors
- Double-sided tape

### Instructions

1. Color each circle in the pattern a different color. Make sure the colors are light enough to allow the labels to show through.

2. Cut out each circle.

3. Begin with the smallest circle. Center it in the next largest circle and adhere.

4. Continue this process with each next largest circle until all the circles are taped together, with the largest being the last one.

5. Now, flip your big circle over and color the other side of it to "look" like Earth.

6. You now have a model of Earth's structure to help you learn which layer is where.

Additional Notes

## Activity for Advanced Students: Make a 3-D Model of Earth

### Materials

- Modeling clay or play dough (recipe in Appendix E) in five different colors
- Dental floss
- Camera

### Instructions

1. Using modeling clay or play dough, form a ball about 1 inch in diameter. This is the inner core.

2. For each of the remaining layers, you will use a different color. For each additional layer, roll out your color of choice to about 1/4 inch thick. You can roll out a circular shape or a square. The amount of play dough needed for each layer will increase as you work toward the outside of your "Earth."

3. Carefully mold the new color around the layers you already have. Take care not to mash it too tightly, as your colors will mix.

4. Continue adding layers until you have a color for each of the layers we have covered in this lesson.

**Additional Notes**

5. Using dental floss, slice a wedge out of your "Earth" to reveal all the colors.

6. Make a diagram or key to go along with your model, listing the description of each layer.

7. Take a picture for your science notebook!

# Unit Two
# The Lithosphere

## Unit Timeline

79 Mt. Vesuvius erupts, burying and preserving three Roman cities, including Pompeii.

1700s Modern geology (science of Earth's history, composition, and [illegible] developed by Jean-Étienne Guettard.

1852 Radanath Sikhdar identifies the highest peak [illegible] is later named Mt. Everest.

1915 Alfred Wegener publishes hi[illegible] that the continents are in constant motion, the "continental d[illegible]ory.

1916 Hawaii Volc[illegible] National Park is established.

1935 Charles Richter develops a scale for measuring the strength of earthquakes.

1953 Edmund Hillary reaches the top of Mt. Everest.

1960s The plate tectonics theory is developed.

1980 Mount St. Helens erupts.

## Unit Two Vocabulary

- active volcano
- calcite
- caldera
- carbonic acid
- caverns
- chemical rock
- cinder cone volcano
- clastic rock
- columns
- composite volcano
- coral caves
- crater
- dormant volcano
- earthquakes
- epicenter
- erosion
- eruption
- extinct volcano
- faults
- focus
- geologists
- geology
- igneous rock
- lava
- lava bombs
- magma
- magnitude
- metamorphic rock
- Milne, John
- organic rock
- Pangea
- plate boundaries
- plates
- plate tectonics
- Richter scale
- rock cycle
- rocks
- San Andreas Fault
- sea caves
- sediment
- sedimentary rock
- seismogram
- seismograph
- seismology
- shield volcano
- soluble
- speleothems
- spelunking
- stalactites
- stalagmites
- sulfuric acid
- vent
- volcanic caves
- volcano
- volcanologist

## Materials Needed for This Unit

- Science notebook
- 4 photocopies of the "Checking It Out" form in Appendix A
- Photocopy of the world map in Appendix A
- Photocopy of the "Speleothem Observation Chart" in Lesson 4
- 1 egg (per student)
- Pan for boiling egg
- Salt
- 12-inch length of cotton yarn
- 2 pint-sized jars
- Saucer
- Epsom salts
- Mixing bowl
- 2 paper clips
- Sand
- Large bucket
- Cookie sheet or tray
- Vinyl tablecloth (to protect table or countertop)
- Small piece of cardboard or plywood
- Narrow-neck bottle, such as a16-ounce plastic soft drink bottle, clean and dry
- Modeling clay or dirt, sand, rocks, etc.
- Small funnel
- 3 to 4 tablespoons baking soda
- 3 to 4 drops liquid dishwashing detergent
- 3 to 4 drops red or orange food coloring
- ½ cup vinegar
- Camera
- Student atlas
- Plastic or wooden building blocks
- Cookie sheet
- Ingredients for Chocolate Nut Layer Bars (See the recipe in Appendix E.)
- Materials for Unit Wrap-Up (See page 73.)

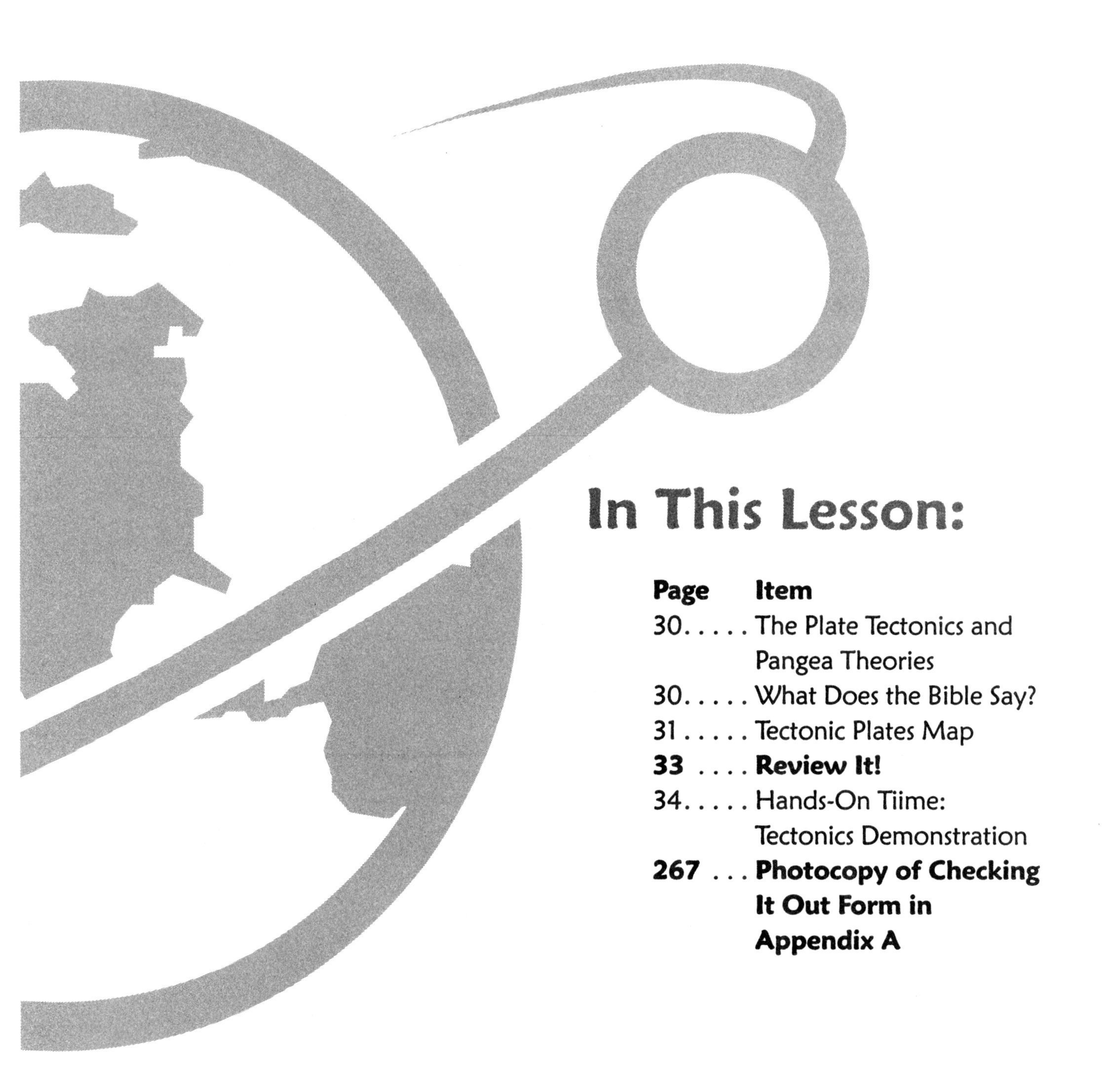

# In This Lesson:

# Lesson 3

# PLATE TECTONICS

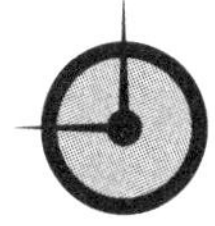

## Teaching Time:

## The Shifting Earth

When God created the universe, including Earth, He created an incredible masterpiece. First, our Earth is beautiful. Second, it is very detailed, right down to the tiniest cell, and everything works perfectly together. God's creation was exactly perfect until Adam and Eve sinned in the Garden of Eden (see the Book of Genesis). This is very important to understand because some scientists disagree tremendously on how things in our world came to be the way they are. Furthermore, not even all Christians agree. Because of these disagreements, not all books you read will teach things in the same way. As we study, we will use the Bible as our guide. We will cover some topics in this unit that are a source of disagreement, but let's remember that scientists are learning more about the world every day. No one but God knows all the answers about our Earth. We will do well to put our trust in God and His Word rather than in man.

**Scripture**

Then God said, 'Let the waters under the heavens be gathered together into one place, and let the dry land appear'; and it was so. (Genesis 1:9)

## The Plate Tectonics and Pangea Theories

**Plate tectonics** (tek **tahn** iks) is one topic about which some scientists disagree. We will explore some things about plate tectonics because it affects other matters we need to study. The basic concept of plate tectonics says that Earth's crust is made up of several pieces called **plates** that fit together a bit like a puzzle. In other words, the crust does not appear to be one solid piece but rather, several pieces that are in constant motion. You can look at the "Tectonic Plates" map in this lesson to see where the different plates are.

Modern science teaches that as these plates shift they sometimes collide and force the surface of the land to change. Many believe this is how mountains are formed. The plate tectonics theory also says that all the continents were once joined together in one supercontinent called **Pangea**. The shifting plates are believed by many to have caused Pangea to break apart, forming several continents. Modern science says that this happened over millions and millions of years ago and that the continents moved very slowly to their present locations. Of course, many Christians have a problem with this concept. Many Christians do not believe Earth is nearly that old.

So then, what do we do with these teachings? Who is right, and who is wrong? Do the plate tectonics and Pangea theories disagree with the Bible?

## What Does the Bible Say?

The Bible says that Earth was created in six days and that God rested on the seventh day. It also says that "Then God said, 'Let the waters under the heavens be gathered together into one place, and let the dry land appear . . . .' " (Genesis 1:9) This Scripture does allow room for the idea of Pangea. Many scientists who are Christians agree that the worldwide flood described in the Bible could have caused the separation of the continents. In this case, the world would not have to be millions and

### Name It!

**plate tectonics**
A theory that says Earth's crust is made up of several pieces called plates that are in constant motion and sometimes collide, forcing changes on Earth's surface.

**plates**
In the theory of plate tectonics, large, thin, relatively rigid layers in Earth's crust that constantly move in relation to one another.

**Pangea**
A theory that says all of Earth's continents were once joined together and have moved apart over time and with the shifting of plates.

### Discovery Zone

What Is a Theory?
Scientifically, a theory is a hypothesis (educated guess) that has survived much experimental testing and has been upheld with a significant amount of data. If this theory then survives generations of testing and remains consistent, the theory can become scientific law. This is all called the Scientific Method.

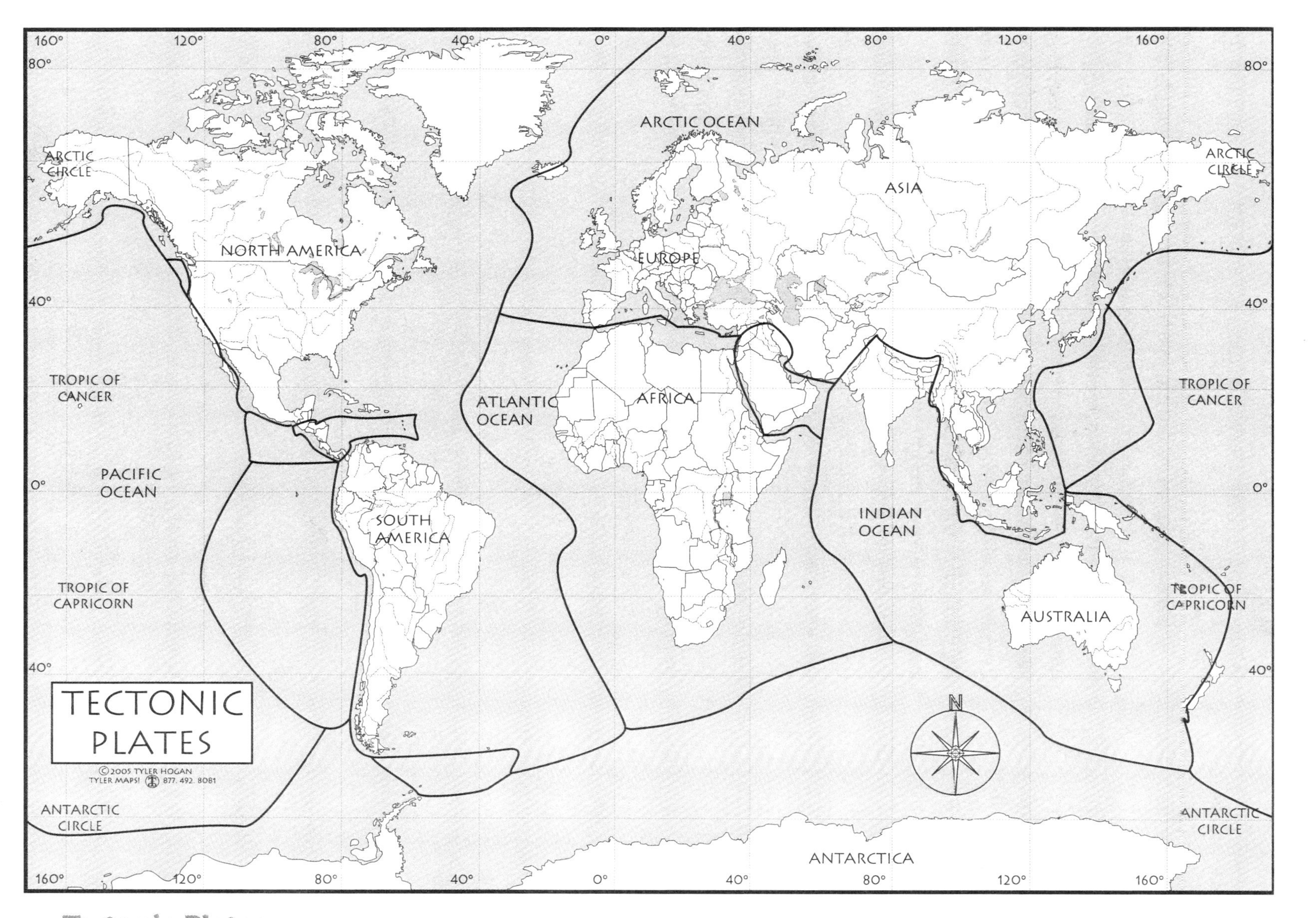


Tectonic Plates

**Discovery Zone**
Learn more about plate tectonics and earthquakes at The Southern California Integrated GPS Network Education Module: http://scign.jpl.nasa.gov/learn

**Discovery Zone**
For more information about plate tectonics and the Bible, check out www.answersingenesis.org

millions of years old. Such a flood could have caused the continents to move apart very quickly. This still allows room for the Pangea theory.

There is a lot of scientific evidence to suggest that Earth's continents were once all one supercontinent and that tectonic plates exist. For instance, modern measurements actually show minor change in the distances between the continents. Also, fossils, rocks, and animal life found on the coasts of some of the continents match up with those that, theoretically, were once joined together. Furthermore, studies of the ocean floor show markings and changes that fit in with the theory.

I cannot tell you for a fact whether Earth's crust is actually a series of tectonic plates or not. I cannot tell you for a fact that all the continents were once joined together in a mass called Pangea. I will tell you that Scripture leaves the possibility open for discussion. For your studies, you need to be aware that the theories exist. You might also like to know that they don't necessarily disagree with the Bible.

## Review It!

*May be done orally or written.*

1. What part of Earth's structure is believed to consist of tectonic plates? ________________

2. The theory that all the continents were once one super-continent is called ________________

3. Name three things that suggest the Pangea theory and the plate tectonics theory might be true.

________________________________________________

________________________________________________

________________________________________________

### Additional Notes

Additional Notes

## Hands-On Time:
# Tectonics Demonstration

## Featured Activity

### Objective
To demonstrate the concept of plate tectonics.

### Materials
- 1 egg per student
- Pan and water for boiling eggs
- 1 teaspoon salt
- Photocopy of "Checking It Out" form in Appendix A

### Instructions
*Adult supervision required.*

1. Place eggs in pan and cover with water.
2. Add salt to the water.
3. Place pan on stove and heat until boiling.
4. Boil for 10 minutes.
5. Have your teacher remove pan from stove and pour off hot water.
6. Refill pan with cold water.

7. Allow eggs to sit in cold water until cool enough to handle.

8. While eggs are cooling, review the lesson on plate tectonics.

9. When eggs can be handled, gently roll each egg on the countertop, cracking the shell.

10. Slide the shell around the egg, allowing the shell to split apart in places.

11. Observe the action. This is a very basic way to view the concept of plate tectonics.

12. Complete a "Checking It Out" form (found in Appendix A) and place it in your science notebook.

## Additional Notes

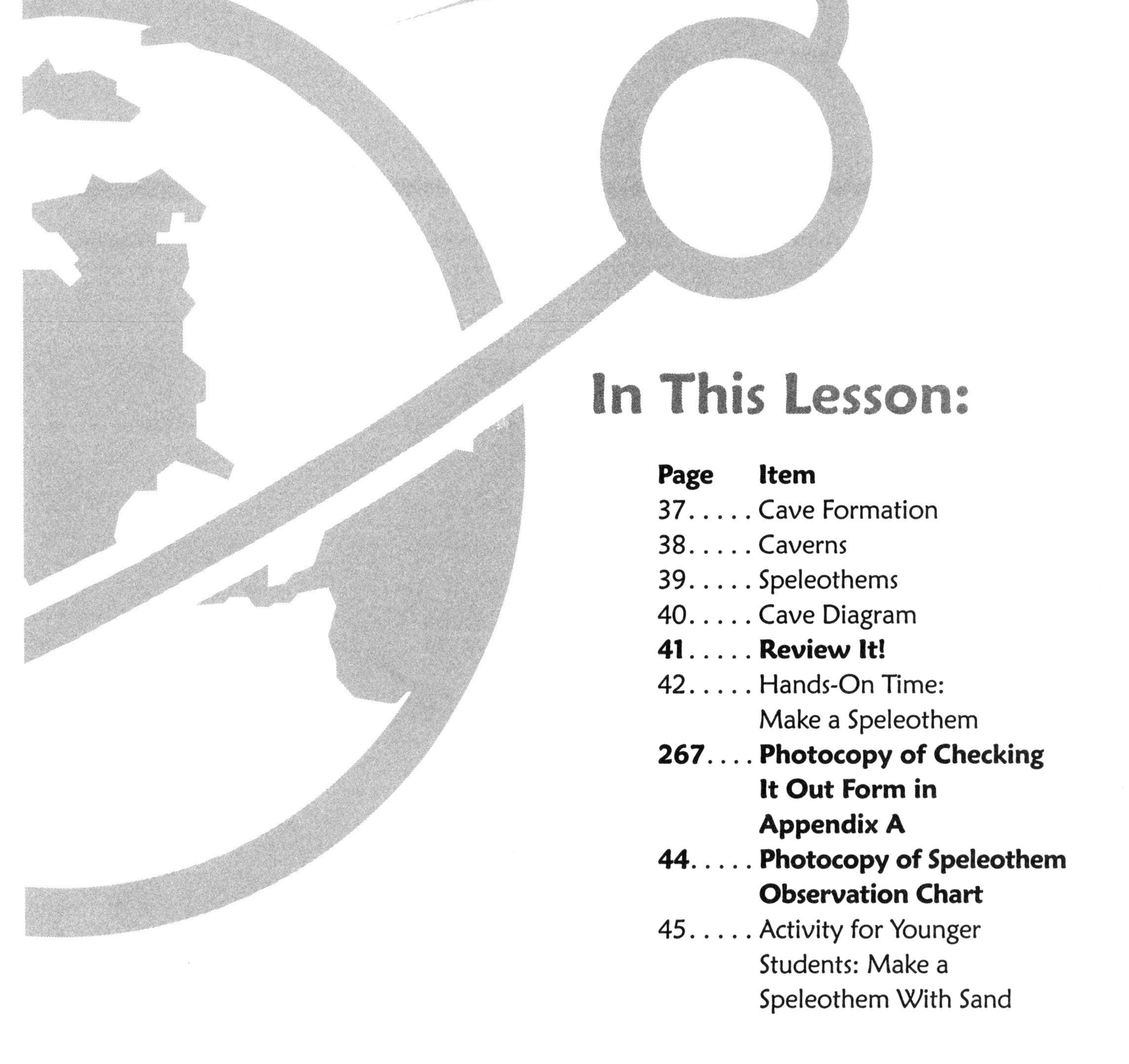

# In This Lesson:

# Lesson 4

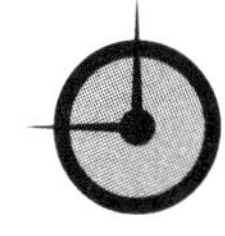

## Teaching Time:

# Spelunking We Will Go!

Have you ever visited a cave? I have, and I loved it! It was dark and damp. There was water seeping down the walls, and there were formations that looked like icicles hanging from the ceiling. Exploring caves as a hobby is called **spelunking**. Spelunking is great fun and can teach you quite a bit about our Earth. I'll admit that when I explored this cave, I had no idea how the cave came to be or what the special features meant. Maybe caves are a mystery to you, too. Let's go spelunking and see what we can learn about God's creation!

## Cave Formation

Do you know that there are many different types of caves? Not only are there different types, but caves are also formed in different ways. Some of the different types of caves are coral caves,

**Scripture**
"So he [Saul] came to the sheepfolds by the road, where there was a cave; and Saul went in to attend to his needs. (David and his men were staying in the recesses of the cave.)" (1 Samuel 24:3)

**Caution!**
Spelunking can be very dangerous unless you have proper equipment, training, and a guide. Do NOT try this on your own!

**Name It!**

**spelunking**
The hobby of cave exploration.

**coral caves**
Caves formed by coral colonies joining each other in shallow waters.

**volcanic caves**
Caves formed by flowing lava and volcanic gases.

**sea caves**
Caves formed when waves erode sea cliffs.

**soluble**
Able to be dissolved.

**caverns**
Caves formed by water and acid that slowly erode soluble rock.

**erosion**
Process of washing away.

**carbonic acid**
Acid formed by rainwater mixing with carbon dioxide in the air; responsible for forming caverns slowly.

**sulfuric acid**
Acid formed by hydrogen sulfide mixing with oxygenated water; stronger than carbonic acid, hence it forms caverns much more quickly.

volcanic caves, sea caves, and caverns. **Coral caves** are formed by coral colonies joining in shallow water. The wind and waves can enlarge the cave by eroding some of the coral. **Volcanic caves** are formed by flowing lava and volcanic gases. Lava tubes are a type of volcanic cave. **Sea caves** are formed when waves erode sea cliffs. We will look more closely at a more familiar type of cave, the cavern.

## Caverns

The surface of Earth is made up of many different types of rocks. Sometimes the rock is soft, other times it is very hard like granite. The softer rocks are often **soluble**, meaning they can be dissolved. Limestone is a common example of a soluble rock. Have you ever dissolved sugar with water? That might give you an idea of what I mean by soluble. Water alone would not be enough to dissolve most rock, however, not even soluble rock like limestone. Water and acids work together to dissolve limestone. **Caverns** are formed by water and acid washing away soluble rock. This process of "washing away" is called **erosion**.

The acid responsible for making caverns usually comes from one of two places. First, there is the acid that is formed by rainwater mixing with carbon dioxide in the air. This forms **carbonic acid**. Carbonic acid is not very strong, so it works slowly. As the rainwater falls, mixing with the carbon dioxide, it soaks into the ground. The acid it forms erodes the limestone, making a hole. Gradually, the path of the rain through the ground forms a type of underground stream, called groundwater, which erodes more and more of the limestone. Eventually, a cavern is formed, and it usually has many winding passages.

Another type of acid that can form caverns is **sulfuric acid**. One way sulfuric acid is formed is when hydrogen sulfide, a gas formed from decomposing leaves and other organic materials, mixes with water that is oxygenated. In the same way as before, limestone is eroded by this mixture and caverns are formed. Because sulfuric acid is much stronger than carbonic acid, caverns are created much more quickly by sulfuric acid than by carbonic acid.

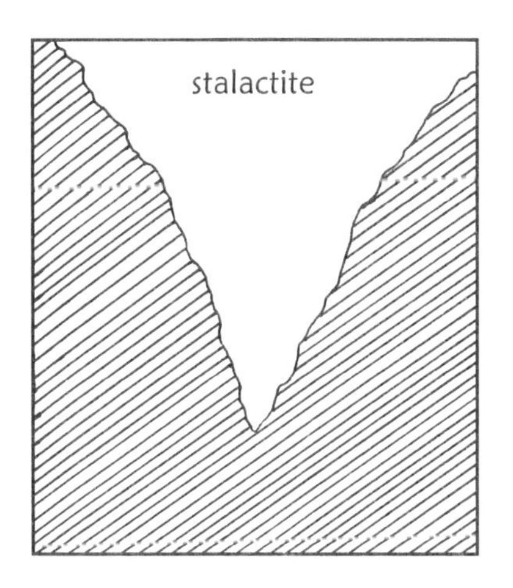


## Speleothems

In the first paragraph, I mentioned that I saw things hanging down in the cave that reminded me of icicles. Actually, these formations are not icicles at all. They are **stalactites** (stah **lack** tites). There are also similar features coming up out of the floor of the cave, and they are called **stalagmites** (stah **lag** mites). When stalactites and stalagmites grow together, they form **columns**. All of these are part of a group of formations called **speleothems**. The word *speleothem* comes from the Greek words *spelaion*, which means "cave," and *thema*, which means "deposit." There are other types of speleothems, but these are the most common. Look at the illustrations in this lesson to see some examples of speleothems.

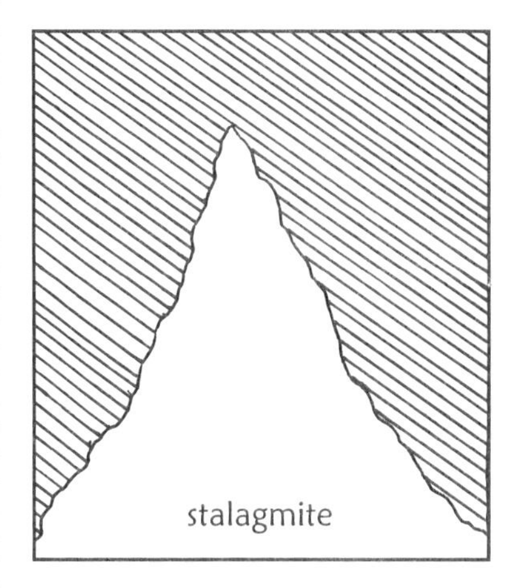


*Note*: To help remember the difference between stalactites and stalagmites, remember that stala**c**tites grow from the *ceiling* and stala**g**mites grow from the *ground*.

Speleothems are cave formations made from minerals that can no longer be held in the water. The name of the mineral deposited is **calcite**. Calcite is a result of the carbonic acid mixing with the dissolved limestone. As water drips through the ceiling of the cavern, it can land on the floor depositing crystallized calcite, which builds up to form a stalagmite. It can also begin to crystallize on the ceiling and form downward, making a stalactite. Now you should be able to understand why the word *speleothem* so perfectly describes these formations. They are indeed cave deposits.

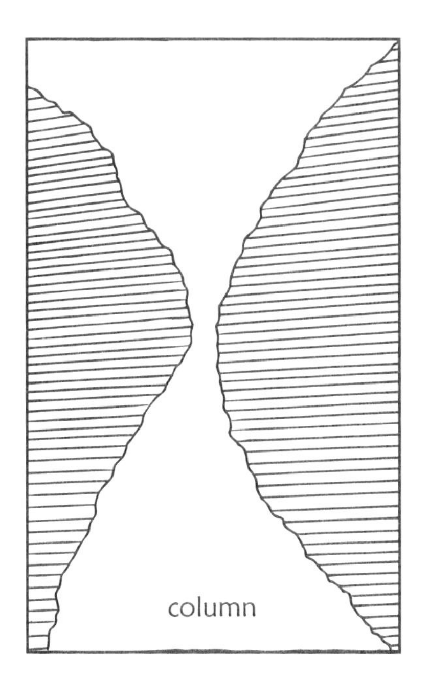


All in all, I hope that you see how fascinating caves are. The type of cave we learned the most about today is the cavern. There are many places in the United States where you can visit a cavern and see all the interesting things we talked about. Maybe your family can make plans to see one on a vacation, unless one just happens to be nearby. If so, you can see if spelunking is a hobby for you!

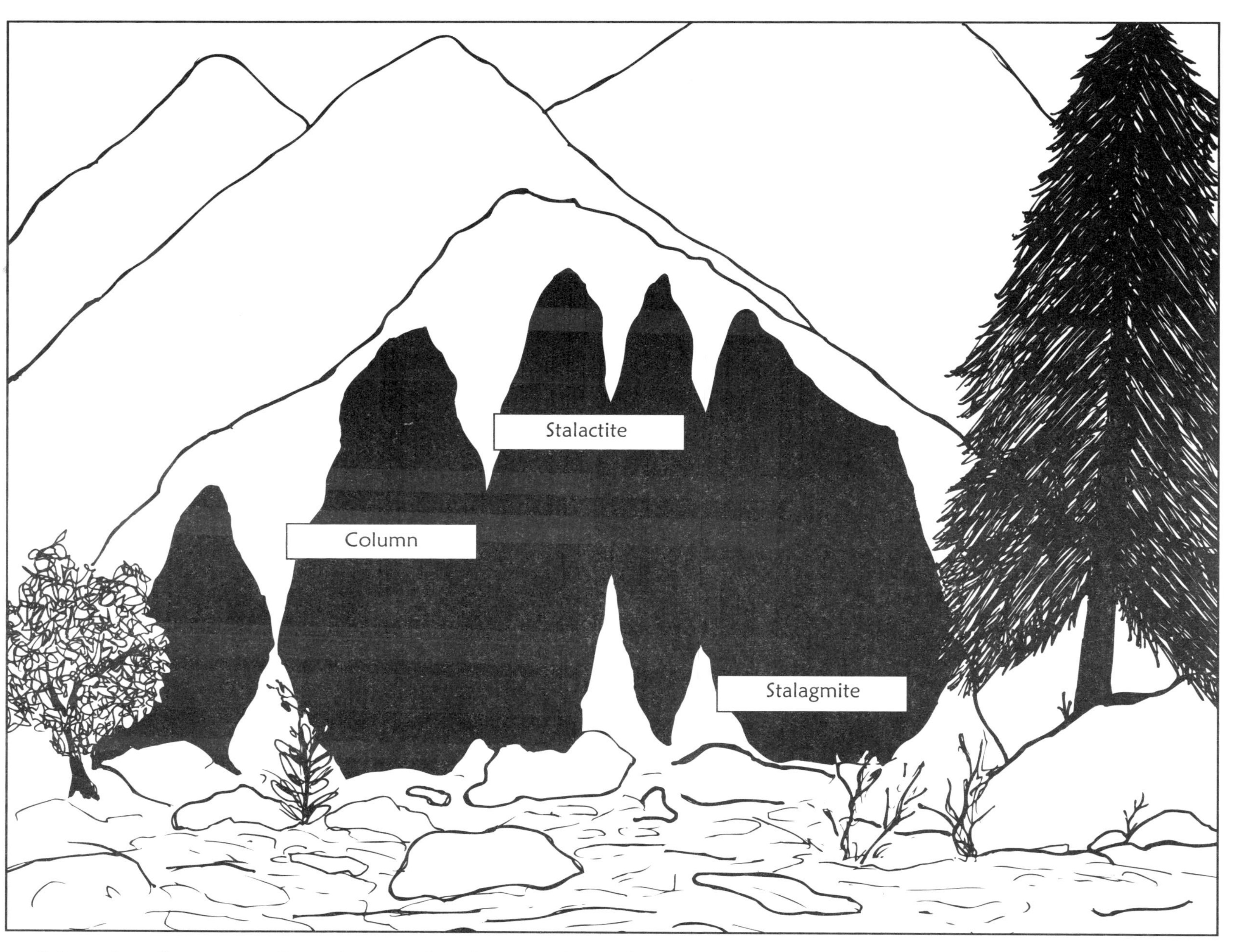


**Mouth of a cave showing stalactite, stalagmite, and column formations**

# Review It!

*May be done orally or written.*

1. What is a speleothem?

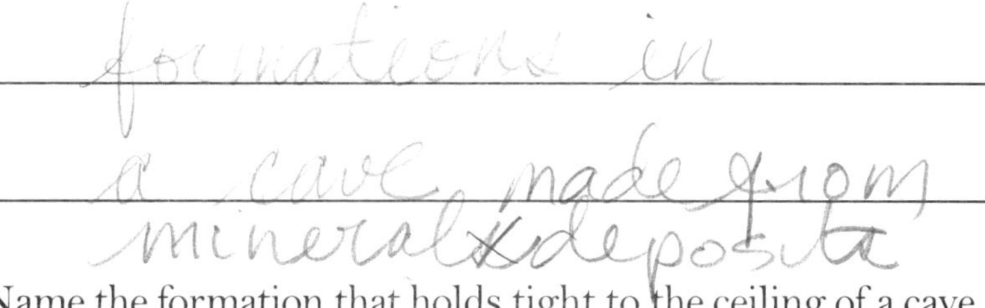


2. Name the formation that holds tight to the ceiling of a cave.

3. What two types of acids erode limestone, forming many caves and caverns?

4. Which type of acid erodes limestone more quickly?

## Name It!

**stalactites**
Cave formations that hang from the ceilings of caves; made from mineral deposits.

**stalagmites**
Cave formations that build up from the ground or floor of caves; made from mineral deposits.

**columns**
Cave formations formed when stalagmites and stalactites grow together.

**speleothems**
Cave formations of various shapes and sizes made from mineral deposits.

**calcite**
Mineral that results from carbonic acid mixing with dissolved limestone, deposited by water and responsible for many speleothems.

Additional Notes

# Hands-On Time: Make a Speleothem

## Featured Activity

*Note*: Your speleothem will take several days to form and will require daily observation.

### Objective

To see how dripping water can deposit minerals and make a formation.

### Materials

- 21-inch length of **cotton** yarn
- 2 pint-sized jars
- Saucer
- 4–5 cups Epsom salts
- 4 cups of warm, distilled water
- Mixing bowl
- 2 paper clips
- Food coloring
- Photocopy of "Speleothem Observation Chart" in this lesson
- Photocopy of "Checking It Out" form in Appendix A

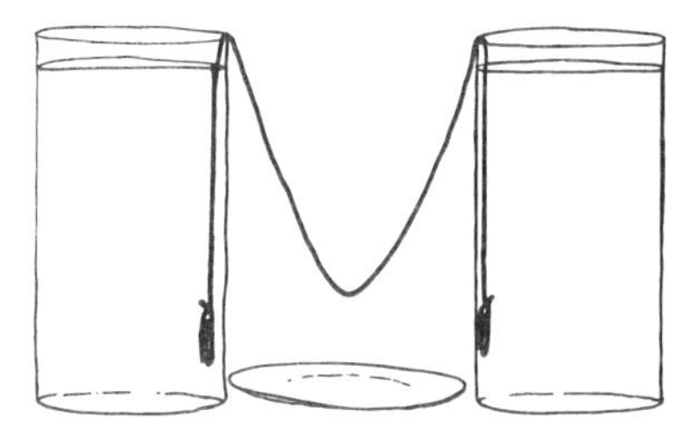

## Instructions

1. Pour 4 cups of very warm water into a mixing bowl, with a drop or 2 of food coloring.

2. Add Epsom salts, stirring continuously. Keep adding salts until no more can be dissolved. You may need to rewarm the water to dissolve the salts completely.

3. Divide this solution evenly between the two jars and place in a sunny spot.

4. Place the saucer between the two jars. Add a small amount of Epsom salts to the center of the saucer.

5. Tie a paper clip onto each end of the yarn.

6. Soak the yarn in the solution, saturating it.

7. Place each end of the yarn in a jar so that the yarn is connecting the two jars. (*See the accompanying illustration.*)

8. Make sure the yarn is inside the solution in each jar.

9. Tug slightly on the center of the yarn so that the center dips down lower than the water levels in the jars.

10. Leave for several days. Each day, observe the project and record your observations in the "Speleothem Observation Chart" that follows.

11. Complete a "Checking It Out" form and put it in your science notebook.

## Additional Notes

## Speleothem Observation Chart

| | Day 1 | Day 2 | Day 3 | Day 4 | Day 5 | Day 6 | Day 7 | Day 8 | Day 9 | Day 10 |
|---|---|---|---|---|---|---|---|---|---|---|
| Water Level (in inches) | | | | | | | | | | |
| Formation on Saucer? Yes/No | | | | | | | | | | |
| Changes in Formation? Yes/No | | | | | | | | | | |
| Sketch of Formation | | | | | | | | | | |

# Activity for Younger Students: Make a Speleothem With Sand

Additional Notes

## Materials

- Sand
- Large bucket
- Water
- Cookie sheet or tray

## Instructions

1. Begin with a bucket about half full of sand.
2. Add water until you achieve a slightly "drippy" consistency.
3. Scoop up some of this mixture in your fist and hold it over the tray.
4. Open your fist slightly to allow the mixture to drip onto the tray. Your formations will vary, but you can see the idea of water and minerals dripping down and creating formations.
5. Have fun!

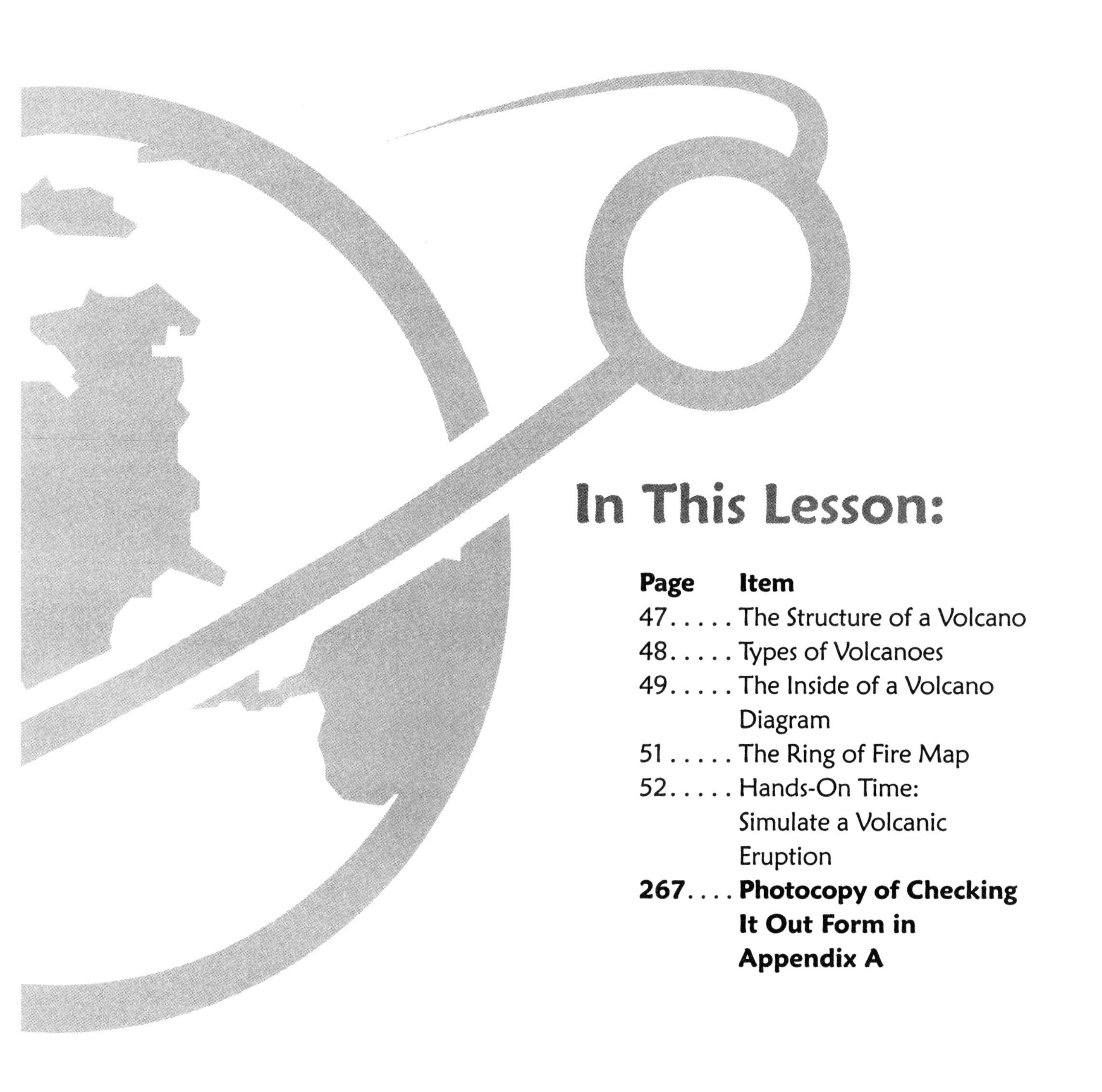

# In This Lesson:

# Lesson 5

# VOLCANOES

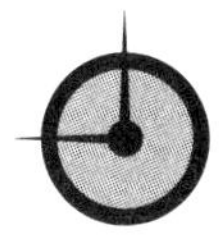

## Teaching Time:

## Hot, Hot, Hot

Do you know what a **volcanologist** is? A volcanologist is a scientist who studies volcanoes. Today, we're going to act like volcanologists and see how much we can learn about volcanoes. Volcanoes are a source of fascination to many people. Their power is more than many of us can comprehend. They are mysterious. Many of us have never visited a volcano and know only what we've seen in pictures or on television. Let's see if we can learn some basic, but interesting, facts about volcanoes and perhaps unravel some of the mystery.

**Scripture**

"Like one who takes away a garment in cold weather, and like vinegar on soda, is one who sings songs to a heavy heart." (Proverbs 25:20)

**Name It!**

**volcanologist**

Scientist who studies volcanoes.

### The Structure of a Volcano

First, do you know what a volcano actually is? Can you define it? Most of you probably have some sort of an idea or picture in your mind. I would like you to write out, tell your teacher, or draw a picture of what you think a volcano is. Now, did your description include words like *hot, boiling, mountain,* or *fire*—or did

### Name It!

**volcano**
Slope or mountain formed by magma.

**magma**
Combination of molten rock, gas, ash, and rock from inside Earth's crust.

**vent**
Tube that connects the inside of a volcano to the deeper part of Earth's crust and the molten rock inside.

**crater**
Sunken area inside a volcano where magma collects.

**eruption**
A reaction caused by pressure building up inside Earth, forcing magma up through a volcano's vent and out of the volcano.

**lava**
Magma that has been forced out of a volcano's vent to the outside of the volcano.

**lava bombs**
Giant clouds of volcanic dust forced out of a volcano by a massive explosion.

**active volcano**
An erupting volcano.

**dormant volcano**
Volcano with long periods of inactivity. (sleeping)

**extinct volcano**
Volcano that no longer erupts. (dead)

your picture illustrate these words? More than likely it did. A **volcano** is a slope or mountain formed by **magma**. Magma is a combination of molten rock, gas, ash, and rock from inside Earth's crust. A **vent**, or tube, connects the inside of the formation to the deeper part of the crust. The top of a volcano has a **crater**. A crater is a sunken area where the magma can collect. As pressure inside Earth builds up, magma is forced up through the vent and can rise to the crater. With enough energy, the magma can be forced out. This is called an **eruption**. If the magma spills out, it is called **lava**. The cooled lava hardens. Repeated eruptions cause a buildup of lava that eventually forms the land formations we know as volcanoes. This is not the only way volcanoes erupt, however. Sometimes there is a massive explosion that releases giant clouds of dust called **lava bombs**. Other times, the walls of a volcano can collapse in. Either way a volcano erupts, you can be sure it's hot!

Notice the diagram of a volcano in this lesson. Can you see the path magma would take as it rises and forces its way out of the volcano?

## Types of Volcanoes

Volcanoes can be **active, dormant**, or **extinct**. Volcanoes are described as active when they are erupting. They are dormant when they have a long period of inactivity. Extinct volcanoes no longer erupt.

As we have talked, have you been picturing a volcano in your mind? You might be picturing a mountain with a hole in the top and that would make perfect sense because many volcanoes look just that way. However, that is not the only form volcanoes can take. As a matter of fact, volcanoes come in several different shapes and vary in size. Let's discuss the different categories of volcanoes. First, there are two categories that are basically in a cone shape, like we were just describing. These are called **cinder cone volcanoes** and **composite volcanoes**. Both types are usually the result of explosive eruptions. However, they are different

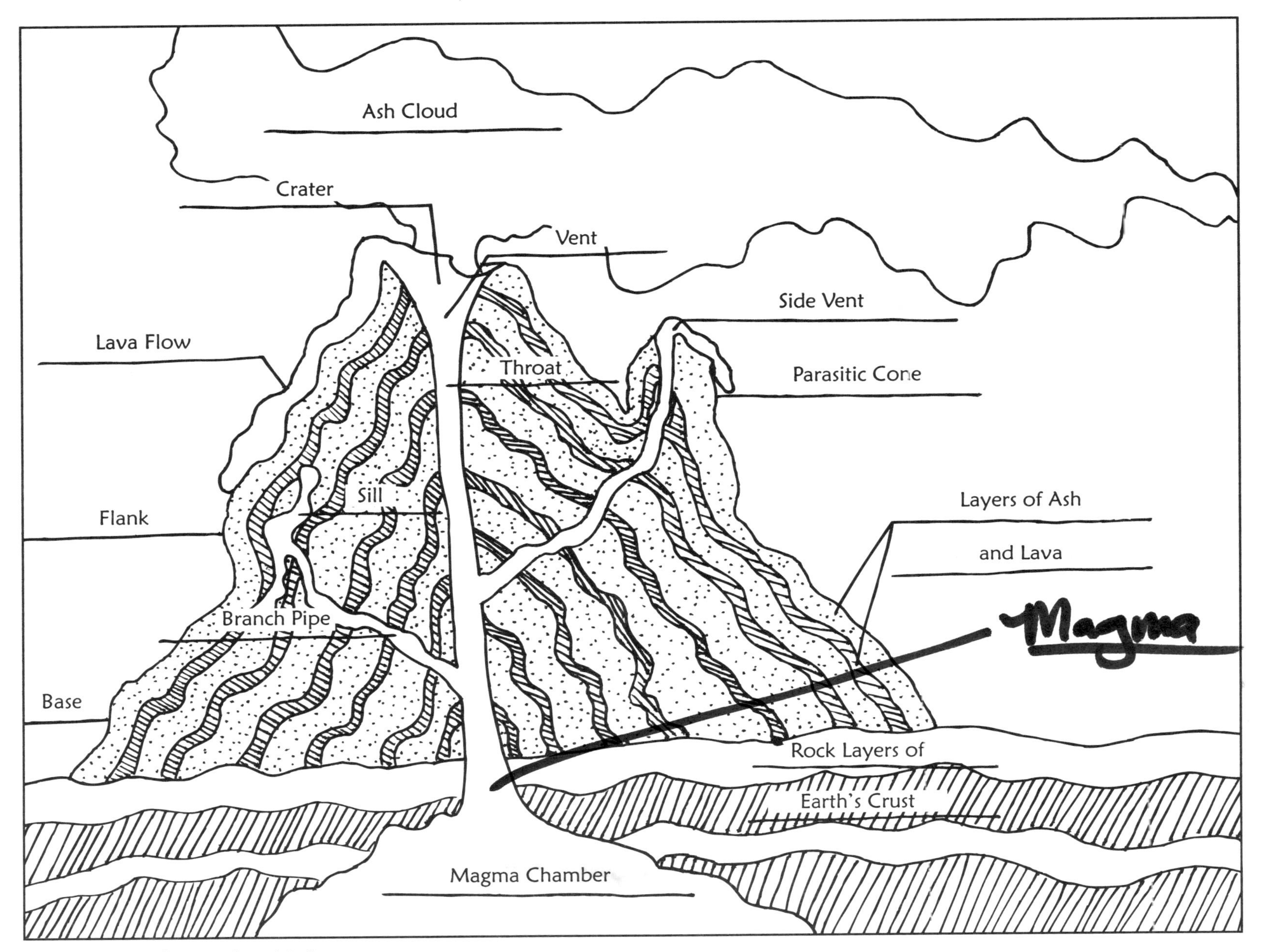


The Inside of a Volcano

**Name It!**

**cinder cone volcano**
Volcano caused by an explosive eruption, cone-shaped and relatively small in size.

**composite volcano**
Volcano caused by an explosive eruption, cone-shaped and large in size due to repeated eruptions over a long period of time.

**shield volcano**
Nonexplosive volcanic mound with gentle, sloping sides, known to vary greatly in size.

**caldera**
Volcano formed low to the ground, caused by the ground's collapsing after an explosive eruption.

in size and makeup from each other. Cinder cones are almost always smaller than composite volcanoes. One reason is that cinder cones are not known to be in existence as long as composites usually are. Composite volcanoes usually have repeated eruptions over a long period of time. The material that comes out of the volcano builds up, increasing the overall size of the volcano. Mount St. Helens in the state of Washington is a good example of a composite volcano.

Next, there are **shield volcanoes**. Shield volcanoes are mounds with gentle, sloping sides. They are nonexplosive. The lava usually travels far from the vent, rather like pouring syrup on a pancake, especially if the syrup is hot. Shield volcanoes can be large or small. The largest shield volcanoes are much larger than even the very largest composite volcano. Two examples of shield volcanoes are found in Hawaii; they are called Mauna Loa (**maw** na **lo** ah) and Kilauea (kill oo **ay** ah).

A fourth type of volcano is the **caldera**. Calderas are round and lower to the ground than the other types. They form when there is an explosive eruption and the ground collapses.

There is a geographical area called "The Ring of Fire" that surrounds the Pacific Ocean. This area is famous for its many volcanoes. Look at the map provided in this lesson to see this hot spot. The coastal areas of all the countries are lined with volcanoes. Many earthquakes happen here as well.

Today we have really covered a lot of ground, so to speak. I hope you have learned some things about volcanoes. Mostly, though, I hope you have developed a greater interest in volcanoes. They truly are fascinating and give me an idea of how powerful God must be if He could create something as powerful as a volcano.

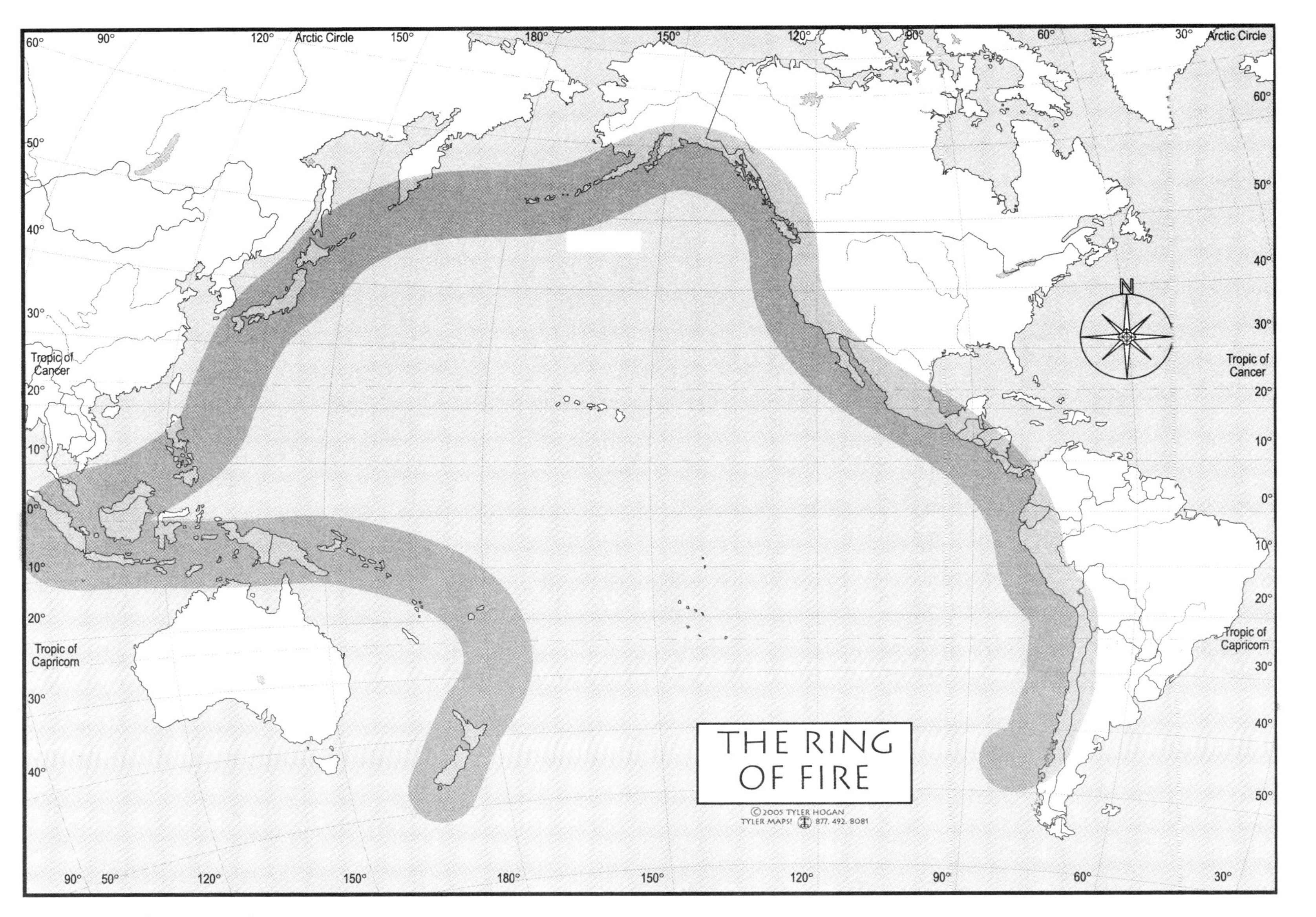


The Ring of Fire

Additional Notes

## Hands-On Time:
# Simulate a Volcanic Eruption

## Featured Activity

*Note to Teacher*: This experiment really shows more about chemistry than volcanoes. However, it's fun and memorable. To make it more applicable, you can try to make your volcano look like a specific volcano, and have your student read about that particular one.

### Objective

To visualize "lava" spewing out of a "volcano."

*Note*: If you make a mound around your bottle with modeling clay or play dough, you will need to allow a day or two for it to harden before performing the demonstration.

### Materials

- Vinyl tablecloth (to protect table or countertop from the eruption!)
- Small piece of cardboard or plywood
- Narrow-neck bottle, such as a16-ounce plastic soft drink bottle, clean and dry
- Modeling clay or dirt, sand, rocks, etc.
- Small funnel

- 3 to 4 tablespoons baking soda
- 3 to 4 drops liquid dishwashing detergent
- 3 to 4 drops red or orange food coloring
- ½ cup water
- ½ cup vinegar
- Photocopy of "Checking It Out" form in Appendix A
- Camera

## Instructions

*Adult supervision required.*

1. Cover your table or countertop with the tablecloth, then place the bottle on a piece of cardboard or plywood. This is a good outdoor project as it is very messy.
2. Cover the bottle with modeling clay, or mounded dirt or sand, creating the shape of a volcano. (Do not cover the mouth of the bottle.)
3. Use a funnel and add 3 to 4 tablespoons of baking soda to the bottle.
4. Add detergent to the bottle.
5. Slowly add water to the bottle.
6. In a separate container, mix the food coloring and vinegar.
7. Slowly add the vinegar mixture to the bottle. The chemical reaction should cause your volcano to erupt.

**Additional Notes**

Additional Notes

8. In your science notebook, record your demonstration and what happened. You can use a "Checking It Out" form (found in Appendix A). Add this to your science notebook. If you take a picture, add it to your science notebook as well.

9. Now, consider again the Bible verse: Proverbs 25:20. Having completed your demonstration, what do you think this Bible verse means? Write out the verse in your science notebook, then also write out your answer to this question.

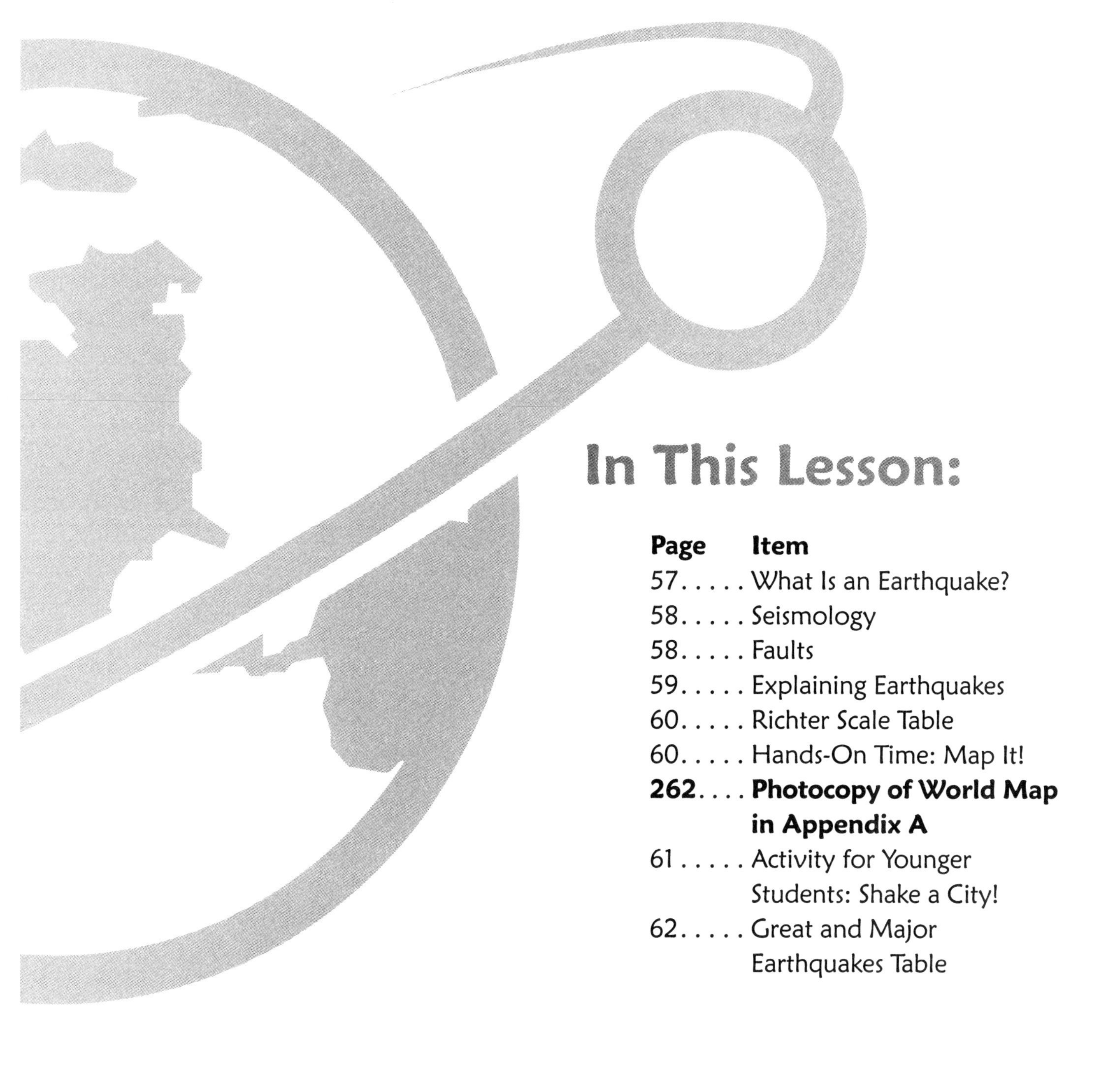

# In This Lesson:

Lesson 6

# EARTHQUAKES

Teaching Time:

## Shake, Rattle, and Roll

Have you ever felt the ground shake? I'm sure that if you have, you still remember it. I grew up in a place where earthquakes aren't very common. One day, however, we did have an earthquake. It wasn't a very strong earthquake; nothing was broken. Still, I remember the coffee cups shaking, as well as the pictures on the wall. I was 7 years old and even though I'm much older now, I haven't forgotten.

**Scripture**

"Then, behold, the veil of the temple was torn in two from top to bottom; and the earth quaked, and the rocks were split . . . " (Matthew 27:51)

### What Is an Earthquake?

It is very possible that you have no idea what an earthquake is. I mentioned feeling the ground shake, and that is what an earthquake seems like. Of course, there is a more scientific definition. **Earthquakes** are the vibrations caused by the shifting of the tectonic plates of Earth's crust. These vibrations can be very strong and destructive or, as in my case, quite mild.

**earthquakes**

Vibrations caused by the shifting of the tectonic plates of Earth's crust; often mild but sometimes very destructive.

**seismology**
The study of earthquakes.

**John Milne**
The scientist who invented the seismograph.

**seismograph**
Instrument that measures and records the shaking, or vibrations, of Earth's crust.

**seismogram**
A drawing, made by a seismograph, of Earth's vibrations.

**faults**
Cracks in the tectonic plates; responsible for many earthquakes.

**San Andreas Fault**
1,000-mile-long fault in Southern California; responsible for many of California's earthquakes.

John Milne's nickname was "Earthquake Milne," and he is known as the father of seismology.

## Seismology

The study of earthquakes is called **seismology** (**size** mall **uh gee**). This word has its root in the Greek word *seismos* (size mos). The origin of this word means "shake." Now that makes sense, doesn't it?

Scientists have worked throughout history to understand earthquakes and, hopefully, to predict them. Predicting earthquakes still proves very challenging. In the early 1890s, a scientist named **John Milne** invented the **seismograph** (**size** muh graf). A seismograph measures and records the shaking of the crust. There were earlier inventions, but this one is considered the most successful. This invention and the improvements made on it since then have enabled scientists to actually see the vibrations of Earth's crust. When the crust vibrates, or shakes, the seismograph creates a drawing of the movement; this drawing is called a **seismogram**. The seismogram will show how strong the earthquake was and how long it lasted.

## Faults

Let's learn about the cause of earthquakes. I've already told you that they are a result of the shifting of the tectonic plates. You may remember learning about plate tectonics a few lessons ago. Earth's crust is believed to consist of about 12 of these plates. They are stiff layers. In other words, they do not bend and flex very easily. Sometimes, as they move about, they develop a crack. These cracks are called **faults**. The faults can move in different directions from time to time, causing earthquakes.

There is a famous fault in the United States. It is called the **San Andreas Fault**. It is located in Southern California. It is over 1,000 miles long and is responsible for the frequent earthquakes in California. Alaska also has a high number of earthquakes. In fact, of the 15 largest quakes in the United States, 10 of them were in Alaska.

Have you ever tried to move something around and had it break instead of bend? That might give you an idea of what happens with the plates. Also, the **plate boundaries**, the places where one plate meets another, are prone to earthquakes. The plates, like faults, can overlap, slide past, or bump each other. This movement can cause earthquakes. In addition, energy can be trapped under these plates, and it must eventually be released. This type of energy, caused by the movement of molten rock under the crust, is also responsible for volcanoes.

**plate boundaries**
Places where one tectonic plate meets another.

**focus**
Spot underground where the central power of an earthquake is located.

**epicenter**
Area on the surface of Earth just above an earthquake's focus.

**Richter scale**
Scale used to rate the strength of an earthquake.

**magnitude**
Measure of strength or quality; an earthquake's magnitude is measured on the Richter scale.

## Explaining Earthquakes

When an earthquake happens, its power is centered on one primary underground spot called the **focus**. The focus can be many miles beneath the surface. On the surface, just above the focus, is the **epicenter**. The vibrations move out from the epicenter in waves and can sometimes be felt far away from the epicenter. The closer to the epicenter you are, the more you will feel the quake. This reminds me of when I toss a rock into the water and watch the rings move out from the spot where the rock hit the water. The waves nearest the epicenter usually cause the most damage. This is because of their strength. There are various scales that scientists use to measure the strength of an earthquake, but the one you will hear about most often is called the **Richter scale**. The numbers associated with this scale are referred to as the **magnitude**. In other words, the magnitude of an earthquake tells us how strong it was. There are over 1,000 earthquakes per year that measure greater than magnitude 5.0 on the Richter scale. Look at the table on the next page to understand more about the Richter scale and the severity of earthquakes.

Additional Notes

**Richter Scale**

| Magnitude | How Strong Is It? | Worldwide Frequency |
|---|---|---|
| More than 8.0 | Great | About 1 per year |
| 7 - 7.9 | Major | 18 per year |
| 6 - 6.9 | Strong | 10 per month |
| 5 - 5.9 | Moderate | 2 per day |

I hope that now you understand a little more about what causes earthquakes. If you're interested in learning more, you can study earthquakes by their damage, by the death toll, and by their location. Any of these will reveal more to you about earthquakes worldwide. If you really love this area of study, you might consider learning more about what a seismologist does.

## Hands-On Time:
# Map It!

## Featured Activity

*Note to Teacher*: Younger students might simply look at an atlas and try to locate the earthquake sites. This will help them with overall map skills and geography. Also, see the activity for younger students that follows.

### Objective

To see where earthquakes have recently occurred.

### Materials

- Student atlas
- Photocopy of the world map in Appendix A
- Pen

### Instructions

Using the chart provided on the following page that lists the great and major earthquakes worldwide for the year 2004:

1. Locate each site in your atlas.

2. Mark each location on the world map. Be sure to include the following:
   - The name of the city and country
   - The date of the earthquake
   - The magnitude

3. Compare your completed map to the map of tectonic plates in Lesson 3. Do the locations relate in any noticeable way to the plate boundaries?

4. File this map in your science notebook.

## Activity for Younger Students: Shake a City!

Using plastic or wooden building blocks, build a "city" on a clean cookie sheet. Pretend that the cookie sheet is the tectonic plate and that it is shifting. Shake the cookie sheet and observe the results. Try making your buildings more secure, perhaps with locking blocks, and shake again. Are there any changes in the results?

### Additional Notes

## Great and Major Earthquakes Around the World - 2004

| Date | Location or Nearby | Magnitude |
|---|---|---|
| Jan. 3 | Southeast of the Loyalty Islands, New Caledonia | 7.1 |
| Feb. 5 | Papua, Indonesia | 7.0 |
| Feb. 7 | Near the south coast of Papua, Indonesia | 7.3 |
| July 15 | Fiji Region | 7.1 |
| July 25 | Southern Sumatra, Indonesia | 7.3 |
| Sept. 5 | Near the south coast of western Honshu, Japan | 7.2 |
| Sept. 5 | Near the south coast of Honshu, Japan | 7.4 |
| Nov. 11 | Kepulauan Alor, Indonesia | 7.5 |
| Nov. 15 | West Coast, Colombia | 7.2 |
| Nov. 22 | West Coast, South Island, New Zealand | 7.1 |
| Nov. 26 | Papua, Indonesia | 7.1 |
| Nov. 28 | Hokkaido, Japan | 7.0 |
| Dec. 23 | Macquarie Island, Antarctica | 8.1 |
| Dec. 26 | Nicobar Islands, India | 7.1 |
| Dec. 26 | West Coast, N. Sumatra, Indonesia | 9.0 |

Source: National Earthquake Information Center, U.S. Geological Survey, as reported in Infoplease. © 2000–2005, Pearson Education, publishing as Infoplease. (http://www.infoplease.com/ipa/A0920923.html) This chart uses moment magnitude, the very newest measure of magnitude. According to the Wikipedia Encyclopedia, "estimates of moment magnitude roughly agree with estimates using other scales such as the Richter magnitude scale." (http://en.wikipedia.org/wiki/Moment_magnitude_scale)

For information on the damage caused by these earthquakes, especially the devastating 9.0 earthquake of December 26, see http://earthquake.usgs.gov/eqcenter/eqarchives/significant/sig_2004.php.

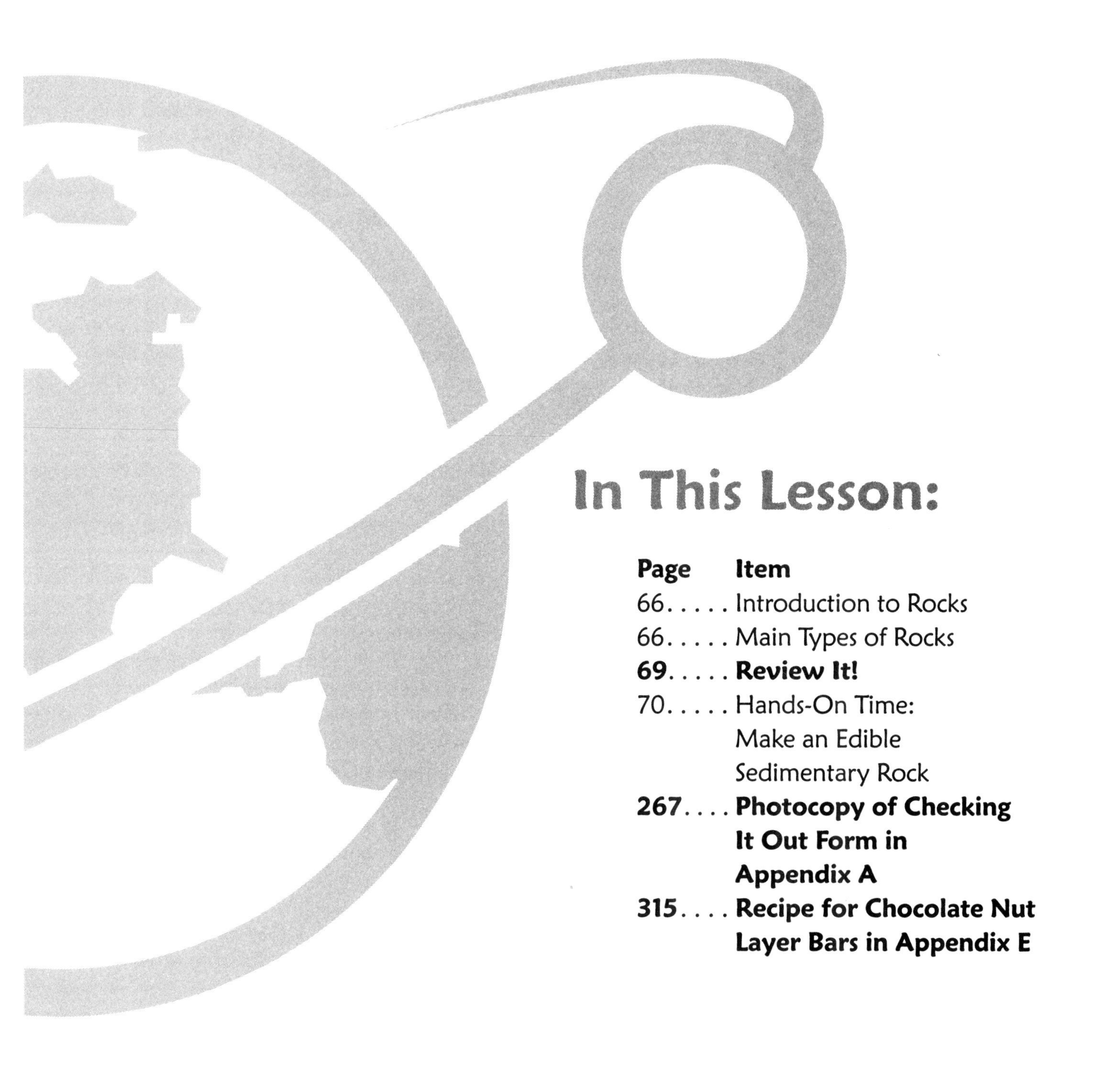

# In This Lesson:

Lesson 7

# ROCK TYPES ON EARTH

## Teaching Time:
## Rocky Ground

As we start out today, I can't help but think of all the things we've studied so far. I don't know if you've noticed or not, but there is a lot to learn about the lithosphere. I'm sure you remember that the lithosphere is just the outer surface of Earth. It is made up of the upper mantle and the crust. All the things we are studying in this unit have to do with the lithosphere. Remember, the parts of Earth that lie under the lithosphere are very deep and very hot. It is difficult to study these aspects of Earth. However, the lithosphere is more accessible. Therefore, scientists have been able to study it more easily through the years and have provided us with many facts. I wonder what scientists will be able to discover in future years about the inner parts of our Earth. There is so much more to discover about this fascinating planet called Earth!

**Scripture**

"Some fell on stony places, where they did not have much earth; and they immediately sprang up because they had no depth of earth." (Matthew 13:5)

**rocks**
Natural mineral formations that make up most of Earth's crust.

**igneous rock**
A type of rock made under the lithosphere when old rocks melt, form magma, cool, and become solid again.

**rock cycle**
The process by which rocks are formed, altered, destroyed, and reformed by geological processes and which is recurrent, returning to a starting point.

# Introduction to Rocks

Have you ever been to a place where there were absolutely no rocks? Now, I'm not talking about being in the city. I mean, have you ever walked on land and seen no rocks? Perhaps, if the land was covered with snow or grass, you didn't see rocks. However, almost anyplace you go, you're bound to find rocks of one type or another. **Rocks** are natural, mineral formations. They make up most of Earth's crust.

I used to live in a town called Stone Mountain, Georgia. Can you suppose what is there? Exactly! There is a *stone* mountain. It is an actual mountain that is all granite. (Granite is a type of rock.) There are trees and dirt on this mountain, but all in all, it is a giant rock. Living in the town near this mountain, I learned that there can be large sections of rock buried underground and that they can make digging very difficult. In this case, the base of Stone Mountain extends for miles underground all around the mountain. I'm telling you this so you can believe that rocks can affect our lives. Even more importantly, rocks can teach us about Earth and help us discover more about where we live. They provide clues to things like volcanic eruptions, the presence of water or floods, and more.

# Main Types of Rocks

There are three main categories of rocks. Within these categories are more categories, but all rocks can be classified, or grouped, in one of the three main ways.

The first type of rocks we'll look at is **igneous** (**ig** nee us) **rock**. Igneous rocks are made under the lithosphere when old rocks melt, form magma, cool, and become solid again. This is part of the **rock cycle**. If the magma cools underground, large crystals form as part of the rock. However, if the magma reaches the surface of Earth (it is then called lava), the rocks will have small crystals. This happens due to how quickly or how slowly the magma cools. Large crystals are formed by slowly cooled magma. Small crystals are formed by magma that cools quickly.

Sometimes, it cools so quickly that glass is formed. Put on your thinking cap. Based on what I just told you, which cools more quickly—magma that comes to the surface of Earth as lava **or** magma that cools inside Earth? Lava forms the smaller crystals, and smaller crystals form by quick cooling. Therefore, we know that lava cools more quickly than magma. Two main types of igneous rocks are granite and basalt.

A second type of rock is **sedimentary** (sed ih **men** tuh ree) **rock**. Sedimentary rocks are formed when loose **sediment**, such as sand and mud deposited from moving water, builds up in layers and hardens. The weight of layers of sediment packs the sediment together. The weight combines with the water and the minerals, causing something like cement to form and binding the elements together into rock. There are three main types of sedimentary rocks: clastic rock, chemical rock, and organic rock.

Sedimentary rock formed from sand and mud in water is called **clastic rock**. Sandstone and shale are clastic rocks. Sedimentary rock formed by water evaporating is **chemical rock**. Examples of chemical rocks are gypsum and halite. **Organic rock** is a type of sedimentary rock formed by plants or animals decomposing and cementing in water or muddy areas. Limestone is an organic sedimentary rock, as is coal. Both of these often contain fossils.

The last main type of rock is **metamorphic** (met uh **more** fick) **rock**. When rocks experience high temperatures and pressure, their structure can change. Sometimes they melt and form new rocks known as the igneous rocks we studied already. However, sometimes they don't melt, but their structure still changes. This is what is called a metamorphic rock. These rocks are studied according to their size, shape, and mineral content. By looking at these features, scientists can learn how they formed and in so doing, can learn more about our Earth.

Scientists who study rocks in this way are called **geologists** (gee **all** uh jists). This branch of science is called **geology**. It is not uncommon to hear scientists use geology as a reason to believe in evolution. By the same token, some creationists ignore

### Name It!

**sedimentary rock**
A type of rock formed when loose sand, mud, and gravel deposited from moving water build up in layers and harden together.

**sediment**
Mineral or organic matter deposited by wind, water, or ice.

**clastic rock**
A type of sedimentary rock formed in water.

**chemical rock**
A type of sedimentary rock formed by the evaporation process.

**organic rock**
A type of sedimentary rock formed by plants or animals decomposing in water or mud.

**metamorphic rock**
A type of rock made when an existing rock is exposed to extreme temperature and pressure, causing the rock to change in form.

**geologists**
Scientists who study rocks.

**geology**
The study of Earth's rocks.

Additional Notes

the discoveries of science in this area because they think it's only about evolution. When I hear of a discovery and a conclusion that don't seem to "line up" with the Bible, I ask, "What else could cause this result? What about a worldwide massive flood? Does God's Word explain this?" I may not always agree with the conclusions some scientists reach, but I consider the evidence and look to God and His Word to help me understand it.

## Review It!

*May be done orally or written.*

1. What is the name of the branch of science that studies rocks and minerals? geology
2. Name the three types of rocks.
   igneous
   metamorphic
   sedimentary
3. Name three types of sedimentary rocks.
   clastic
   chemical
   organic
4. Which type of rock is made from cooling magma?
   igneous
5. Review the rock cycle.

**Additional Notes**

Additional Notes

## Hands-On Time:
# Make an Edible Sedimentary Rock

## Featured Activity

### Objective
To visualize the layers of sediment that combine to make one rock.

### Materials
- See recipe in Appendix E for Chocolate Nut Layer Bars. (No peanuts used.)
- Photocopy of the "Checking It Out" form in Appendix A

### Instructions
1. Bake the cookie bars according to the directions.
2. Allow bars to cool in pan.
3. Carefully cut the bars into rectangles.
4. Observe the layers of the bar.
5. How many layers do you see?
6. How is this similar to sedimentary rock?
7. How is it different? (Think about the role water plays.)
8. Complete a "Checking It Out" form, including your responses to the above questions.

# Unit Three
# The Hydrosphere

## Unit Timeline

1840 Louis Agassiz proposes that glaciers once covered central Europe.

1872–1875 The British ship HMS *Challenger*, an oceanographic research ship, discovers peaks and valleys on the ocean floor, much like those on dry land.

1969 Crafts designed for underwater laboratories allow scientists to begin staying underwater for long periods of time.

1974 The Safe Drinking Water Act ensures that people will have safe water to drink in the United States.

1986 *Alvin*, a submersible watercraft, makes a trip to the *Titanic* remains.

## Unit Three Vocabulary

- alpine glacier
- Antarctica
- aquifer
- calving
- chasms
- continental glacier
- current
- deposit (verb)
- drainage
- equator
- evaporate
- firn
- freshwater
- glacier
- groundwater
- hydrologic cycle
- hydrosphere
- iceberg
- impermeable
- infiltration
- Northern Hemisphere
- oceanology
- permeable
- piedmont glacier
- precipitation
- recharging
- saltwater
- saturated
- Southern Hemisphere
- transpiration
- tropics
- valley glacier
- water table

## Materials Needed for This Unit

- Science notebook
- 3 photocopies of the "Checking It Out" form in Appendix A
- Photocopy of the unlabeled "Hydrologic Cycle" diagram in Appendix A
- Globe
- Crayons
- Sliding board
- Play dough (See the recipe in Appendix E.)
- 12-inch length of string
- Chalk
- Paved area
- 2 large, clear glasses
- Dirt, sand, and gravel (Aquarium gravel would work well.)
- 2 paper cups
- Wooden plank about 2 feet long
- Watch or clock with a second hand
- Materials for Unit Wrap-Up (See page 111.)

OCEAN RESEARCH II
RIVER EXPLORATION
JETSET

# In This Lesson:

# Lesson 8

# INTRODUCTION

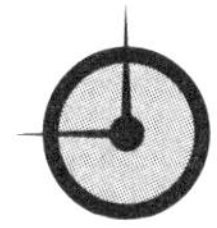

## Teaching Time:

## Water, Water Everywhere

It's time to begin a new unit today. If you are working through the units of this book in order, you have covered some basics about Earth's structure and its creation. You have also studied the lithosphere and some of its features. Now it's time for our studies to take a dive, so to speak. We are going to begin a unit on the **hydrosphere**. The word *hydrosphere* comes from the Greek root *hydro*, which means "water." This is the big hint that we are going to be learning about the water on Earth.

If you have a globe, you should get it out and have a look at it. A globe will give you a good idea of how much of our Earth is covered in water. Can you make a guess as to how much? Do you think water covers more or less than half of Earth? You might be surprised, but water covers over 70 percent of the surface of our Earth. What's more, a globe cannot show you all the

**Scripture**

"In the six hundredth year of Noah's life, in the second month, the seventeenth day of the month, on that day all the fountains of the great deep were broken up, and the windows of heaven were opened. And the rain was on the earth forty days and forty nights." (Genesis 7:11–12)

**Name It!**

**hydrosphere**

The water found on Earth.

**Name It!**

**groundwater**
Water that drains through the soil and is stored underground in aquifers.

**glacier**
Large mass of ice that moves slowly.

**hydrologic cycle**
Process by which water moves from Earth's surface to the sky and back again.

**evaporate**
To change from a liquid to a vapor.

**transpiration**
Process by which moisture in plants or trees turns to a vapor.

water on Earth because some of it is located underground. Wow! And I thought there was a lot of land on Earth!

## Water Sources on Earth

Going back to the globe, you'll most likely think that the largest amount of water on Earth seems to be in the oceans. You're right. Oceans account for 97 percent of Earth's water. Where else do we have water? If you said lakes and rivers, you are right again. Lakes and rivers are important; however, they hold only a very small amount of the water on Earth. The remainder is in **groundwater**, the water that drains through the soil, and in **glaciers**, which are large masses of ice that slowly move. Glaciers form in places where the winter snow never fully melts. The snow packs down, and new rain and snowfall add to the mass over a long period of time, forming a glacier.

During this unit, we will learn much more about the oceans, lakes and rivers, groundwater, and glaciers. For the remainder of this lesson, we will focus on something called the **hydrologic cycle**. This will give us a good foundation for our study of the hydrosphere.

## The Hydrologic Cycle

The hydrologic cycle is the process by which water moves from Earth to the sky and back to Earth again. This is a very simple definition, but it is a very good place to start.

As I've told you already, most of Earth's surface is covered with the water in the oceans. In fact, you learned that oceans cover almost 70 percent of Earth's surface. The Sun causes this water (and other water on Earth) to **evaporate**, or change from water into a vapor. You might be interested to know that only the water evaporates. The salt and other minerals in the water are left behind.

Other moisture that changes to a vapor is the moisture in plants and trees. This process is called **transpiration**, but it is very similar to evaporation.

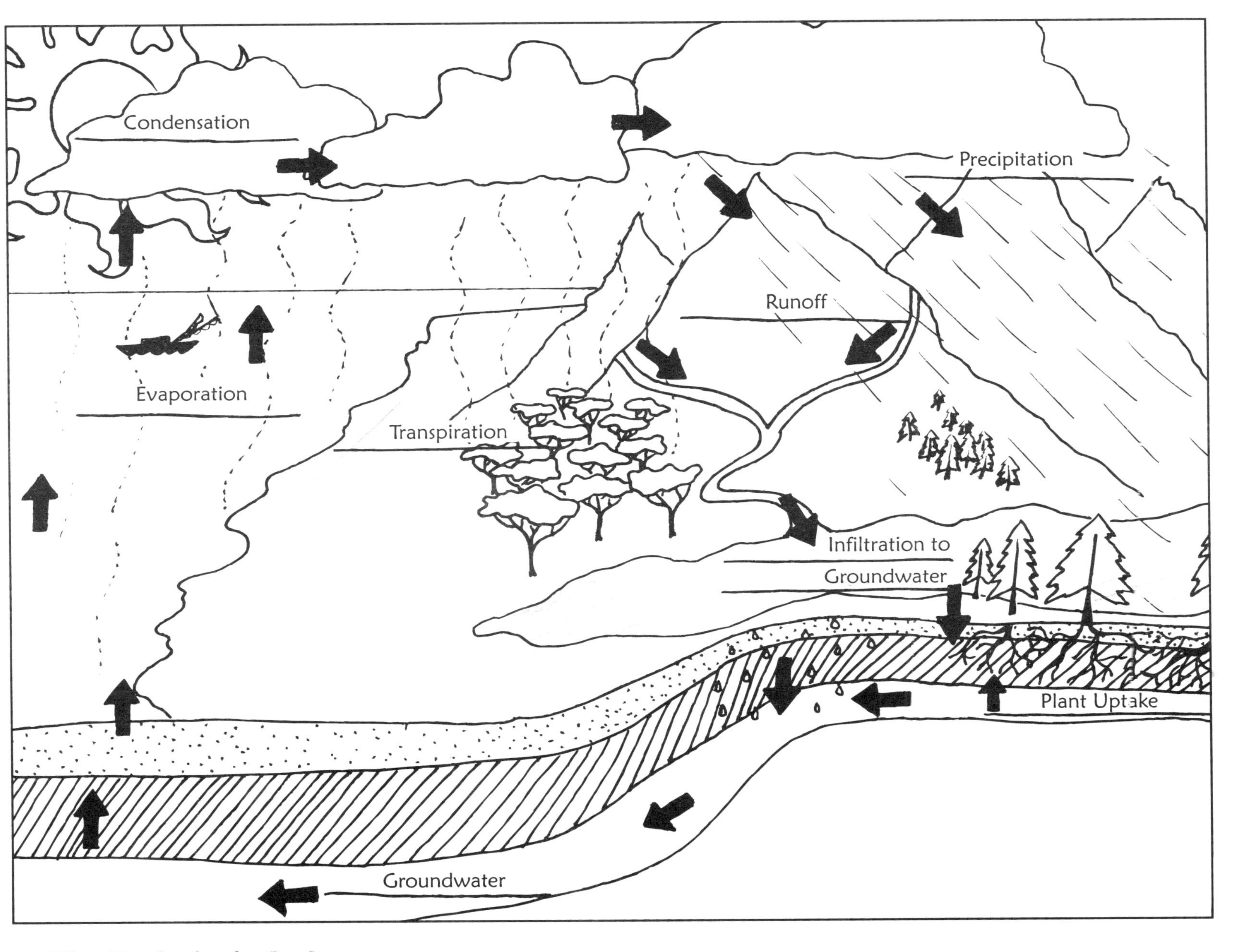


The Hydrologic Cycle

**precipitation**
Water that falls to Earth in the form of rain, snow, sleet, hail, or freezing rain.

**infiltration**
Process by which precipitation soaks through the soil and becomes groundwater.

All this vapor, which is moisture, collects in the sky and eventually forms clouds where the water droplets have come together. Soon there is too much moisture for the air to hold, and **precipitation**—moisture that falls to Earth—forms. Precipitation can come in the form of rain, snow, sleet, or hail. When there is precipitation, the water runs into the lakes, rivers, and oceans. It also soaks through the soil and becomes groundwater. This is called **infiltration**.

This cycle of evaporation, cloud formation, and rain goes on continuously and is known as the hydrologic cycle. I've included a diagram of the hydrologic cycle for you to study. You will learn about this cycle in much more detail in your high school courses, so we will not go any further. For our study, I would like you to understand what evaporation means and know what the hydrologic cycle is. This will help you to understand other things throughout the unit.

# Hands-On Time: The Hydrologic Cycle

## Featured Activity

### Objective

To reinforce the teaching of the hydrologic cycle and assist in learning the vocabulary.

### Materials

- Photocopy of the unlabeled "Hydrologic Cycle" diagram in Appendix A
- Crayons

### Instructions

1. Label the diagram as best as you can, according to the definitions in your vocabulary.
2. Color the diagram as neatly as you can.
3. Add to your science notebook.

## Activity for Younger Students: Act Out "Evaporation"

### Instructions

*Note*: This activity requires a sliding board.

Today you get to act out the evaporation process. To do this, you need to pretend you are water. The ground is your ocean. The top of the slide is the sky.

1. Run around in the "ocean." Say, "I'm the ocean water. I'm saltwater."

2. Begin climbing the slide. Say, "I'm evaporating from the ocean. I left the salt and minerals in the ocean. I'm freshwater now."

3. At the top of the slide, yell, "Condensing into clouds!"

4. Slide down the slide, calling out, "Precipitation in the form of rain." (You can change this to "snow" or "hail.")

5. On the ground you are once again water in the ocean.

6. Repeat the process several times.

7. Draw a picture or write a paragraph for your science notebook.

### Additional Notes

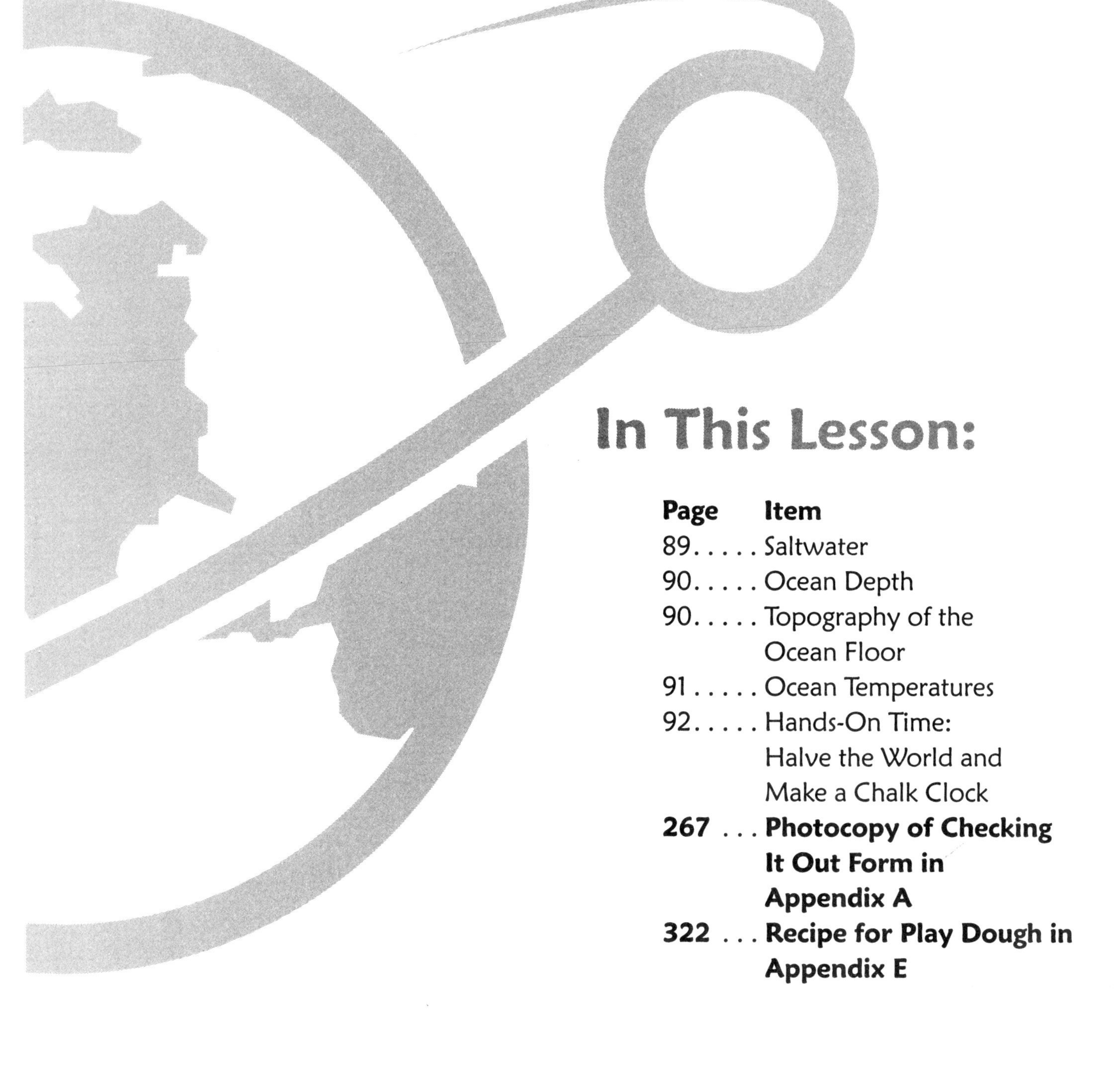

# In This Lesson:

# Lesson 9

# OCEANS

## Teaching Time:

## Pass the Salt, Please

In our last lesson, you learned some basic information about the hydrosphere. You learned that the word *hydrosphere* refers to all the water on Earth. We also learned a little bit about the oceans on Earth. Do you remember how much of the surface of Earth is covered with water? The answer is: over 70 percent. This means that more than half of Earth is covered in water. As we saw in our last lesson, most of that water is in the oceans. In fact, 97 percent of our Earth's water supply is found in the oceans. Today we're going to learn more about the oceans and what they mean to us.

**Scripture**

"Whatever the LORD pleases He does, in heaven and in earth, in the seas and in all deep places." (Psalm 135:6)

### Saltwater

If you've ever been swimming in the ocean, you probably noticed the taste of the water. You may have also felt the water sting your eyes. That happens because the ocean is salty. About 3 percent of the ocean is salt. This may not sound like very

Learn more about the *Challenger*, an early oceanic research ship, at www.coexploration.org/hmschallenger.

**deposit**
As a verb: To put, place, or leave something somewhere; in this case, by a natural process.

**saltwater**
Water found in oceans, seas, and some lakes that contains high levels of salt and other minerals left behind by the evaporation process.

**chasms**
Very deep valleys or canyons, often found in glaciers.

much to you, but it is certainly enough to taste. The salt and other minerals found in the oceans come from one main source and that source is rocks. When it rains, the water falls down on Earth and soaks into the ground. It also falls into lakes and rivers. As the water flows, it eventually makes its way to the ocean. As it travels, it washes away some of the minerals, including salt, from the rocks and soil. All these minerals are **deposited**, or placed, into the ocean water. When the water evaporates, the salt and other minerals are left behind, making the oceans **saltwater**. Saltiness is one interesting fact about oceans, but it's not the only one. Let's learn more!

## Ocean Depth

The average depth of the oceans is 36,080 feet. That is about six miles deep! The deeper you go, the more pressure there is. You could not swim six miles below the surface. The pressure is so great at this depth that it will even destroy submarines. Because of this, most of our ocean depths are totally unexplored. Mankind will probably never know the great mysteries that lie deep in the oceans.

## Topography of the Ocean Floor

Just when you think you've learned all there is to know about the oceans, there's more. Do you know that there are mountains under the surface of the oceans? It's true. There are entire mountain ranges practically hidden under the ocean. There are also volcanoes under the ocean. Some of the mountaintops and volcanoes peek up through the water's surface, forming islands and groups of islands. Hawaii is a good example of this, along with the Aleutians in Alaska. In addition to the mountain ranges, there are great **chasms**, or very deep valleys and canyons, that are deeper than the Grand Canyon.

## Ocean Temperatures

Earth's oceans vary greatly in their temperatures. Let's think about it and see what we can discover. Remember what you've already learned about Earth and its axis. Earth's tilt leaves some parts of the world always warmer than others. The regions around the **equator** (ee **kwate** er) are always fairly warm. These areas are called the tropics. Tropical waters can reach 79 degrees Fahrenheit in the upper few feet of water. However, no matter where the ocean is located, the deeper you go the colder the water. Why? Well, the sunlight can shine down only a certain distance and therefore, beneath that depth it is always dark and quite cold.

One last item to cover in our ocean lesson is ocean **currents**. Do you know what a current is? A current is a flow of water. Ocean currents have patterns of movement. The oceans in the **Northern Hemisphere** (**hem** is fear), or all of Earth located above the equator, flow in a clockwise motion. The oceans in the **Southern Hemisphere**, all of Earth below the equator, flow in a counterclockwise pattern. The continents on Earth interrupt these patterns, but these are the basic patterns of movement.

As I'm sure you can tell, there is a lot to learn about oceans. We have only touched on a few of the very many things there are to know about this subject. There are scientists who devote their whole adult lives to the study of the oceans, which is called **oceanology** (oh shin **all** oh gee).

**equator**
Imaginary line that separates the Northern Hemisphere from the Southern Hemisphere; labeled as the 0 degree line of latitude.

**tropics**
The regions around the equator, characterized by consistently warm temperatures and warm ocean waters.

**current**
A flow of water.

**Northern Hemisphere**
All of Earth above the equator.

**Southern Hemisphere**
All of Earth below the equator.

**oceanology**
The study of Earth's oceans.

Additional Notes

## Hands-On Time:
# Halve the World and Make a Chalk Clock

## Featured Activity

### Objective
To show the Northern and Southern Hemispheres in a memorable way and, also, to understand more about the patterns of the ocean currents.

### Materials
- Play dough (See the recipe in Appendix E.)
- 12-inch length of string
- Chalk
- Paved area
- Photocopy of the "Checking It Out" form in Appendix A

### Instructions
Each of these activities is very simple but teaches important information. The accompanying artwork illustrates one of these activities. Be sure to record your demonstration on a "Checking It Out" form and file it in your science notebook.

1. Make a ball of any size with the play dough. This represents Earth.

2. With your fingernail, make marks to represent the North Pole and the South Pole.

3. Gently turn your Earth on its "side."

4. Use the string to cut through Earth at the halfway point. This is where the equator is located.

5. Now set your South Pole properly on the table. The cut portion should be facing up.

6. Set the other half back in its proper place. The upper portion of Earth, all that is above the equator, is the Northern Hemisphere. The bottom portion is the Southern Hemisphere.

7. Now, take your chalk outside to the paved area for the next portion of your Hands-On Time.

8. Draw a large circle, at least three feet in diameter, on the pavement. (*See illustration.*)

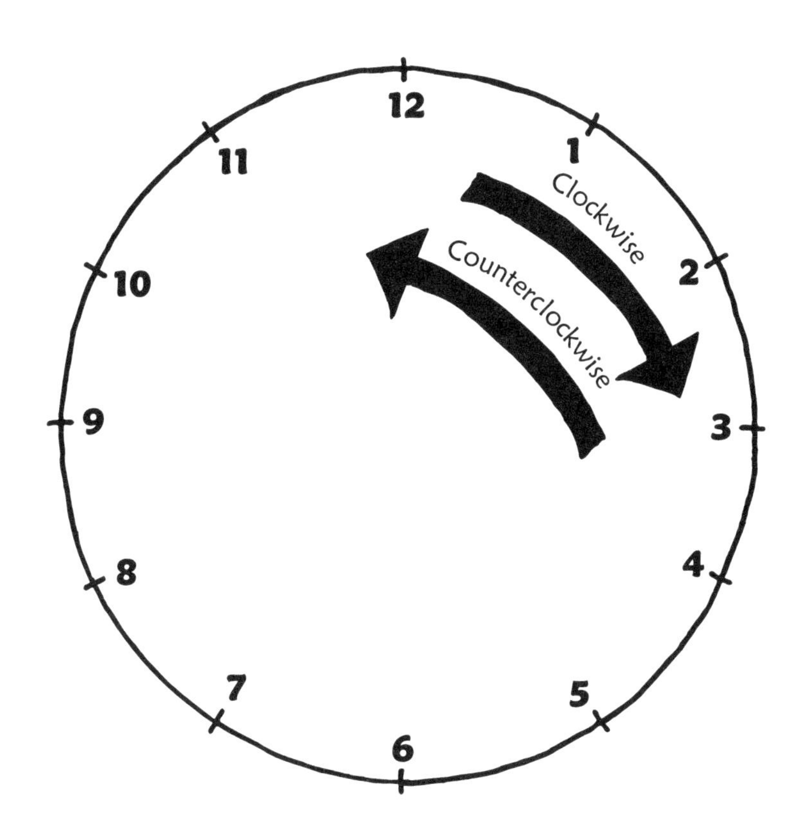


## Additional Notes

Additional Notes

9. Label the circle with numbers like a clock.

10. Stand on the number 12.

11. Walk toward the 3. You are moving in a clockwise direction, or the way a clock moves. This is the main direction of the ocean currents in the Northern Hemisphere.

12. Now walk back toward the 12 from the 3. You are now moving counter, or against, the way a clock moves. You are going "counterclockwise." This is the main direction of the ocean currents in the Southern Hemisphere.

13. Complete a "Checking It Out" form and place it in your science notebook.

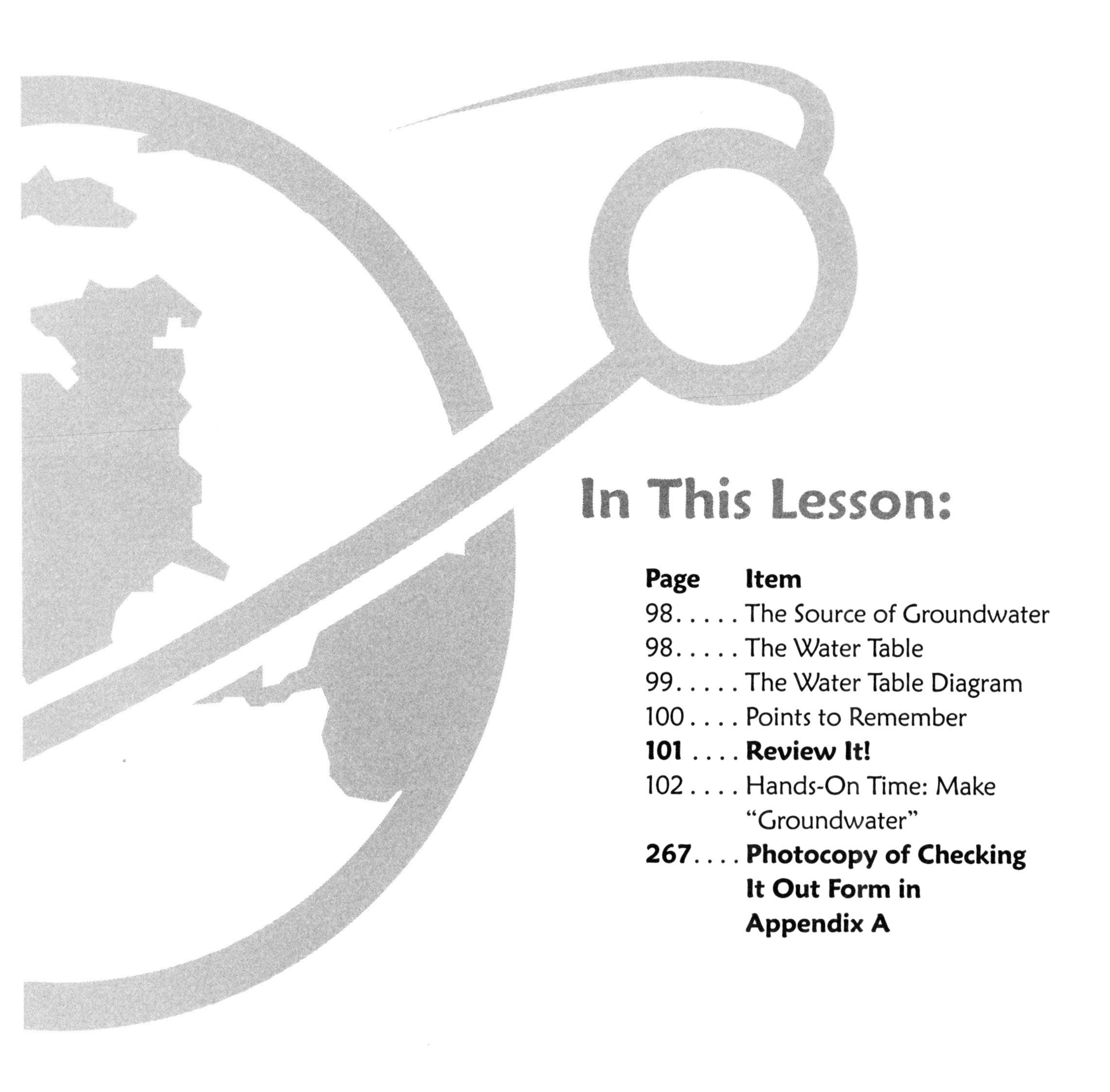

# In This Lesson:

# Lesson 10

# GROUNDWATER

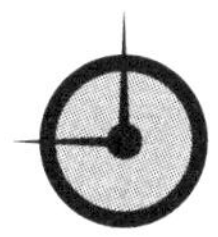

## Teaching Time:

## Let's Go Underground

As you have already learned, Earth has a lot of water. There is saltwater and freshwater. You have also learned about the hydrologic cycle and how only the water is evaporated from the ocean, not the salt and other minerals. This gives us freshwater in the clouds, and that is what falls to Earth in one form of precipitation or another. Do you remember what precipitation is? A hint is in this paragraph. Precipitation is water that has been evaporated and returns to Earth in the form of rain, sleet, hail, or snow. Dew is also a form of precipitation.

**Scripture**

"He also brought streams out of the rock, and caused waters to run down like rivers." (Psalm 78:16)

Our last lesson focused on oceans. Oceans are saltwater. You learned that oceans make up the largest portion of the hydrosphere. You learned that the average ocean depth is about six miles, and you learned that the oceans in the Northern Hemisphere flow in the opposite direction from those in the Southern Hemisphere. It would be a good idea to think back on your last lesson and review.

Today we're going to shift our thinking from the oceans on

**freshwater**
All the water on Earth that's not saltwater; found in lakes, streams, rivers, and ponds, as well as in precipitation.

**groundwater**
Water that drains through the soil and is stored underground in aquifers.

**infiltration**
Process by which precipitation soaks through the soil and becomes groundwater.

**drainage**
The natural or artificial removal of water from a given area.

**aquifer**
Layers of rocks, soil, and sand that contain water.

**saturated**
Holding as much water as possible.

Earth to the sources of **freshwater**. Freshwater is all the water on Earth that is not saltwater. Make sure you know that freshwater is not necessarily clean water. Freshwater is found in several main places on Earth. It is in lakes, rivers, and streams, as well as in groundwater and glaciers. (*Note*: Not all lakes are freshwater. Some lakes are saltwater, not freshwater.) Today we're going to study **groundwater**, which is water that soaks through the ground and is stored there.

## The Source of Groundwater

Where does groundwater actually come from? Basically, it comes from rain and snow. When precipitation occurs, some of it goes directly into bodies of water such as oceans, lakes, and rivers. What about the precipitation that falls onto land? Some of the water runs off the land and ends up in lakes, and other bodies of water, but some of it soaks right into the soil. This is called **infiltration** (in fill **tray** shun). God then uses this water to nurture our plants and trees and grasses. However, there is often more water than the plants need. The water then continues soaking into the ground as far as it can. Some types of soil have better **drainage**; that is, they allow more water to pass through than other soils do.

## The Water Table

You may be trying to figure out what happens next. Well, the water begins to build up in a formation of layers of rocks, soil, and sand that can absorb water. This formation is called an **aquifer** (**ak** wiffer). Groundwater is the water that is stored in an aquifer. Groundwater is held in an aquifer by filling in the cracks and holes in the aquifer. We will look further into this in the Hands-On Time this week.

Eventually, the ground around this area can reach the point where it is holding as much water as it can. That means it is **saturated** (**sa** chur ate ed). When this happens, the water collects,

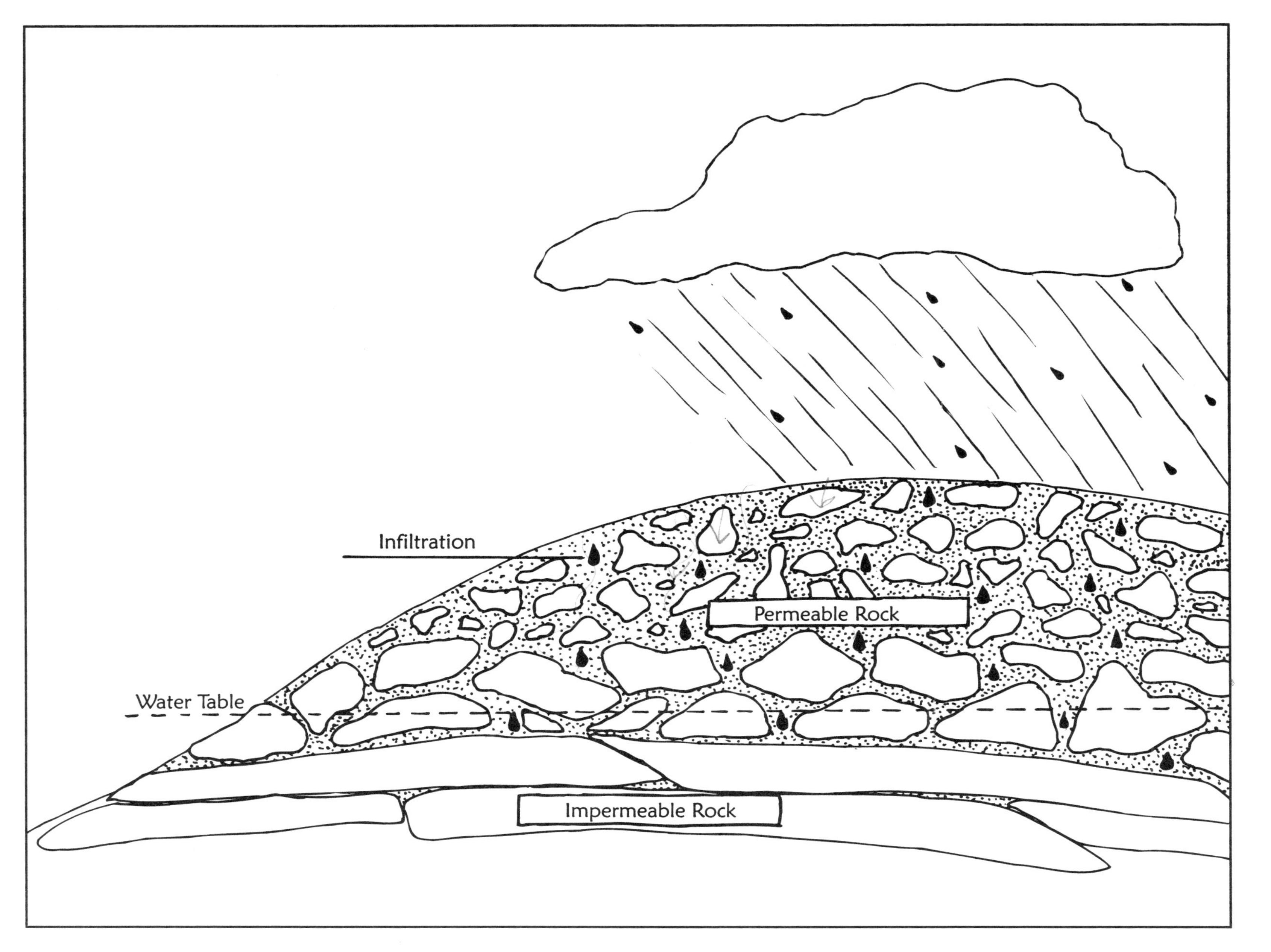


The Water Table

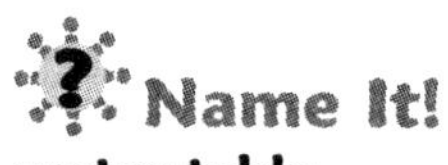

**Name It!**

**water table**
Imaginary line underground, located just above the level of saturation.

**permeable**
Able to hold water or allow water to pass through.

**impermeable**
Cannot hold or be penetrated by water.

**recharging**
The process of precipitation infiltrating the ground and adding to the underground water supply.

forming a stream that flows underground. The area just above where the stream forms is called the **water table**. This is an imaginary line between soil that can hold water, known as **permeable** (**per** me uh bull), and soil that cannot, **impermeable**. You can look at "The Water Table" diagram in this lesson to see this more clearly.

In some locations the water table is deeper than in others. You may wonder why. It all depends on how often it rains in a specific location and what the ground is like. If it rains often, the water table may not be very deep, since the new rain will keep replenishing, or **recharging**, the ground. The soil will become saturated more quickly than in a location that doesn't receive much water. Areas that don't receive much rain can have very deep water tables because there is more time between rainfalls, allowing the soil and plants to dry out.

## Points to Remember

Groundwater may be a little confusing to you. Here are some main things you'll want to remember.

- Groundwater is water that is stored in aquifers underground.
- Groundwater is a large source of freshwater.
- Most of the water used in the United States comes from groundwater.
- Groundwater can become a stream that delivers water back to the oceans, along with salt and minerals it absorbs from the ground.

As you can see, a lot happens after a rainfall or snowfall. Just because we can't see the water, doesn't mean it isn't there. Groundwater is an important part of our lives and another way God provides for us.

## Review It!

*May be done orally or written.*

1. What is the name of the process of water soaking into the soil? infiltration

2. Is groundwater saltwater or freshwater?

3. What is an aquifer? (ak wiffer)
layers of rock, soil & sand that contain water (can absorb water)

4. What does *saturated* mean?
something holding as much water as it can

5. What do we call the imaginary line between permeable and impermeable soil? Water Table

### Additional Notes

Hydrologic Cycle
Water Cycle =
> evaporation
> transpiration
> condensation
> precipitation
> infiltration

Additional Notes

## Hands-On Time:
# Make "Groundwater"

## Featured Activity

### Objective
To see an aquifer and observe saturation and a water table.

### Materials
- 2 large, clear glasses
- Sand
- Gravel (Aquarium gravel would work well.)
- Pitcher of water
- Photocopy of the "Checking It Out" form in Appendix A

### Instructions
(*See the illustration accompanying this activity.*)

#### First Glass

1. In both glasses, create an aquifer by adding sand and gravel in alternating layers until the glasses are three-quarters full.

2. Slowly pour water into one of the filled glasses. Notice the water moving through the cracks along the way.

3. Continue pouring water into the glass until it is full. Make sure the water is above the aquifer.

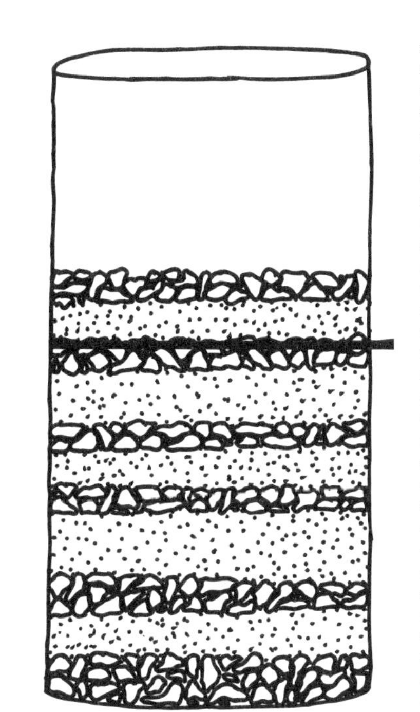

glass 1 glass 2

### Second Glass

4. Slowly pour water into the second glass. Stop about one inch below the top of the aquifer. Notice the level of the water. The area just above the level of water is the water table. Below this point the aquifer is saturated. The glass acts as impermeable rock.

5. Now, "rain" on your "ground" by slowly adding more water to the glass until the water table is about one-half inch below the top of the aquifer. This is recharging the groundwater.

6. Complete a "Checking It Out" form and add it to your science notebook.

### Additional Notes

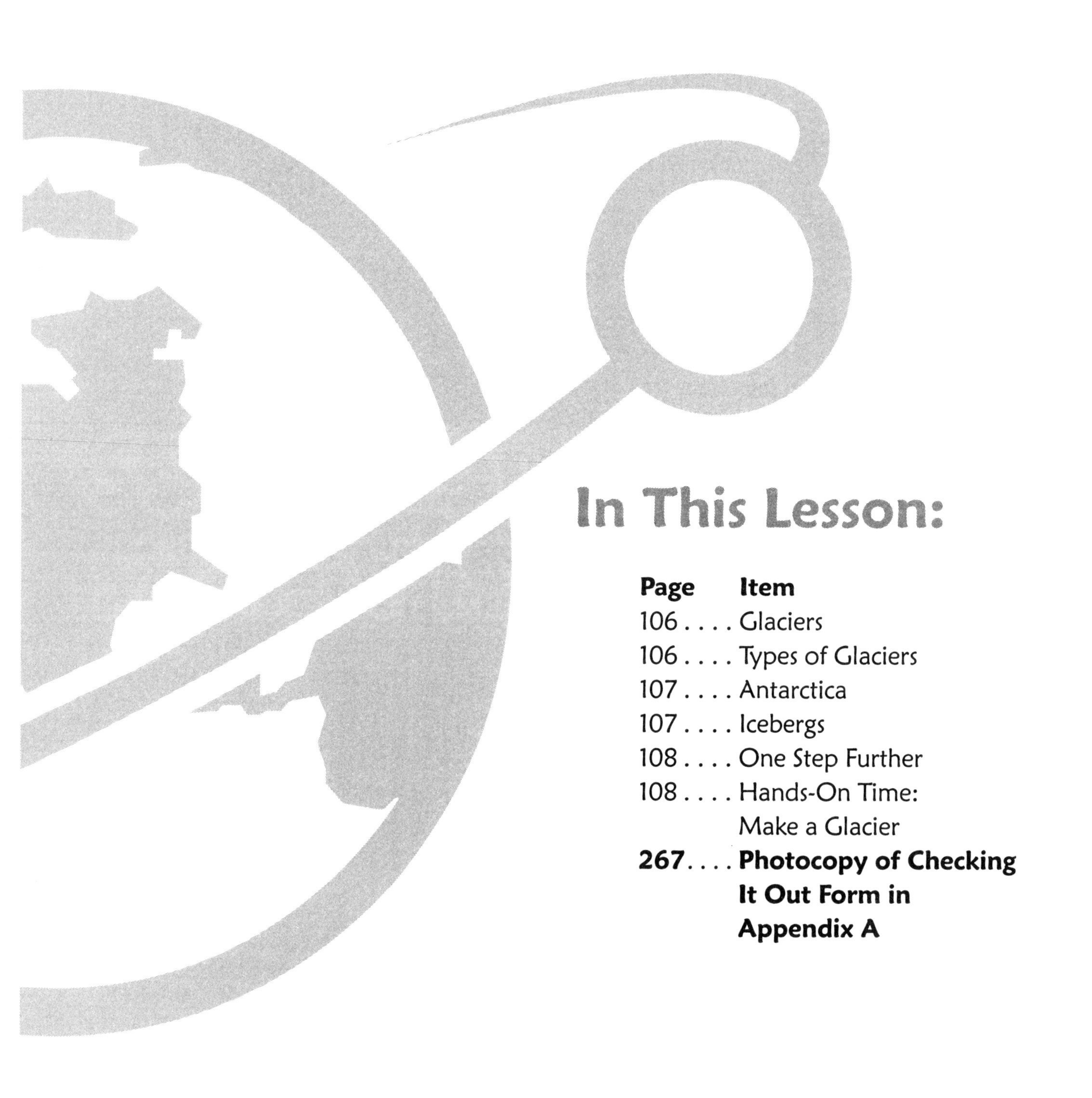

# In This Lesson:

Lesson 11

# GLACIERS AND ICEBERGS

## Teaching Time:
## Iced Down

Wow! You have learned a lot about Earth's hydrosphere. You have learned about saltwater, largely found in the oceans, and freshwater, particularly in the form of groundwater. You may think you know all you need to know about the water on Earth, but then you'd miss out on some fun! Today's lesson is on a source of freshwater that many of us will never see in our lifetimes. I think we're in for an adventure!

Many may think that lakes and rivers are the largest source of freshwater on Earth. I believe this is because we can see lakes and rivers quite easily. The truth is that lakes and rivers contain only a small portion of the freshwater on Earth. Even groundwater is not in first place for having the largest amount of freshwater on Earth. The largest amount of freshwater found on Earth is in our glaciers and icebergs. Today we're going to learn more about what glaciers and icebergs actually are and where they are found.

 **Scripture**
"By the breath of God ice is given, and the broad waters are frozen." (Job 37:10)

### Discovery Zone

Check out many glacier facts at www.factmonster.com. Choose "World & News," then "World Geography."

### Name It!

**glacier**
Large mass of ice that moves slowly.

**firn**
Layer of hard-packed snow on a glacier.

**continental glacier**
Enormous ice sheet that covers most of a continent or large island.

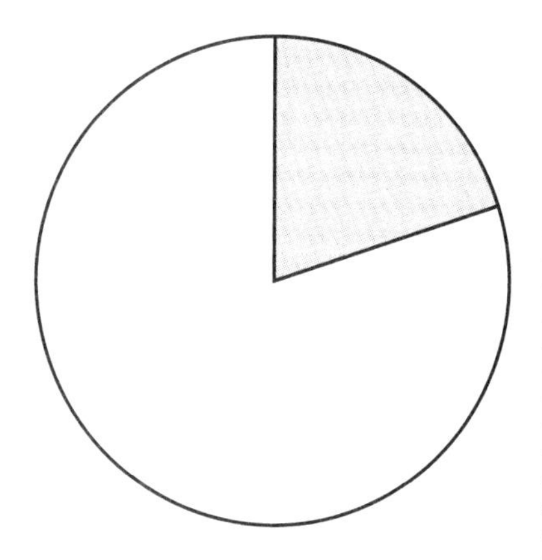

**Legend**
☐ Not covered in ice
☐ Covered in ice

This pie chart illustrates "4/5"—the portion of Greenland that is covered in ice!

## Glaciers

Do you know what a glacier is? A **glacier** is a huge mass of ice that flows, or moves, usually quite slowly. Glaciers form in places where snow falls but never completely melts. More snow falls, packing down the snow that was left from before. A layer of hard-packed snow forms; this is called a **firn**. The more snow that falls, the harder it becomes for the snow to melt. If you live in an area that receives much snow, you may have seen a similar event. As snowplows clear parking lots, they pile the snow all together. Long after the snow is melted on the ground, the piles of snow in the parking lots are still around. In areas where the snow never melts, something similar happens and the snow-packed area grows and grows, getting heavier and heavier, which causes it to begin a very gradual slide. This is a glacier. Glaciers are much larger than the piles of snow in your local parking lots!

## Types of Glaciers

Would you believe that glaciers exist on all our continents except one? Australia is the only continent in the world that has no glaciers. Not all glaciers are the same, however. There are different types of glaciers. One type in particular is called a **continental glacier**. This is an enormous ice sheet that covers most of a continent or a large island. Greenland is a good example of a continental glacier. Over four-fifths of Greenland is covered in ice. The accompanying pie chart illustrates this coverage. At the center of the island, the ice is about 8,000 feet thick. That's more than a mile!

In addition to the continental glacier, there is also the **alpine glacier**. This is the most common type of glacier. The word *alpine* refers to mountains. Based on that, where do you think alpine glaciers can be found? In mountainous areas, of course! Using the information you have already learned about glaciers, you can assume that alpine glaciers form on mountains that have snow that can't fully melt. This could be because the climate in that region is so cold. It could also be because the mountain is so tall that it is always cold enough at the top to keep the snow from fully melting. As the snow packs down, the weight begins to move the ice mass down the slope of the mountain. This type of formation is often seen in river valleys. For this reason, an alpine glacier is also called a **valley glacier**.

## Antarctica

The world's largest glacier is **Antarctica**. It is a continental glacier. In fact, only the eastern region of the entire continent is solid earth. The remaining part of this vast continent is a series of islands joined together by ice. We cannot see the islands because of all the ice. At its thickest point, the ice is about 15,000 feet thick. That's more than three miles!

## Icebergs

Icebergs are formed as a result of glaciers. An **iceberg** is a very large chunk of ice that breaks off a glacier in the melting process. (Remember, I said that glaciers form from snow that doesn't *completely* melt. I did not say that the snow *never* melts.) The formation of icebergs from glaciers is called **calving**. Icebergs are much larger than they usually appear. Much of an iceberg is actually underwater. We see only the very tip of the iceberg.

At the beginning of this lesson, I told you we were going to study a large source of freshwater. You have now learned that this source is glaciers and icebergs. Here's one more fact for you: Glaciers contain over 75 percent of Earth's freshwater supply. Next time you worry or hear that our Earth may run out of water, remember glaciers and icebergs. That doesn't mean you shouldn't conserve water, though. After all, you probably can't run out back to a glacier for more.

### Name It!

**alpine glacier**
Glacier that forms in mountainous areas; also called a valley glacier.

**valley glacier**
Glacier often seen in river valleys, formed by the movement of alpine glaciers down mountain slopes.

**iceberg**
Large chunk of ice that breaks off of a glacier in the melting process.

**calving**
The breaking off and forming of icebergs from glaciers.

**Antarctica**
The continent surrounding the South Pole, almost entirely covered by an ice sheet, making it a continental glacier and the world's largest glacier.

### Discovery Zone

One of the largest icebergs was formed in 1956. It was 208 miles long and 60 miles wide. That is larger than some states. Over 90 percent of the iceberg was under water.

Additional Notes

## One Step Further

For more study, you can:

1. Research a third type of glacier, the **piedmont glacier**.
2. Discover the name for the study of glaciers.
3. Research the *Titanic*'s demise.

## Hands-On Time:
# Make a Glacier

## Featured Activity

(*Note*: This must be prepared early in the day or the night before.)

### Objective:

To simulate the movement of glaciers and observe the debris they can leave behind.

### Materials

- 2 paper cups
- Dirt, sand, gravel
- Wooden plank about 2 feet long
- Watch or clock with a second hand
- Photocopy of the "Checking It Out" form in Appendix A

### Instructions

1. Place "debris" (de **bree**)—dirt, sand, and gravel—in one of the cups until it is one-third to one-half full.

2. Add water to fill the cup.

3. Fill the second cup with water.

4. Place both cups in the freezer overnight or until the water is totally frozen.

5. Place the plank on a stack of books or on an overturned bowl to incline.

6. Remove both cups from the freezer and remove their ice in one piece. These are your "glaciers."

7. Place the glaciers on one end of the sloped plank. (Place them with the largest flat side down.)

8. Time them as they melt and move down the plank. How long does it take each to reach the bottom?

9. Observe the path made by the glacier with the dirt. What do you see?

10. Complete a "Checking It Out" form and place it in your science notebook.

*Note*: When glaciers move along, they carry rocks and dirt along with them. The combination of rocks, dirt, ice, and water along the path erodes the land. Consider your small "glacier." What do you think would be different with a massive glacier on the side of a mountain?

**Additional Notes**

# Unit Three Wrap-Up

Now that you've completed the unit, it's time to wrap up all that you've learned and show what you know!

1. Copy this page and place a check mark in each box for work you've completed.
   - ❑ Lessons read
   - ❑ Daily Reading Sheets
   - ❑ Hands-On Time
   - ❑ Checking It Out forms
   - ❑ Vocabulary
   - ❑ Coloring Page

2. Now that you've accounted for the work up to this point, you should review all that you've studied. Here are some ways to do that.
   - ❑ Create a folderbook that covers the main ideas from each lesson. Instructions can be found in Appendix F.
   - ❑ Write a composition that reviews each lesson. See "Write About It!"—the composition worksheet and form in Appendix A.

     *(Note: If your science notebook is complete and you've done a good job with your Daily Reading Sheets, it should be fairly easy to get the facts you need for these summaries.)*

3. Show What You Know!—Complete the quiz on the next page to the best of your ability. After your teacher has reviewed your work, go back and use your book to answer anything you didn't know the first time through.

# Show What You Know!

Answer as many questions as you can without using your book or notes. You get **10,000** points for each correct response. After going through the quiz once with your book closed, have your teacher review your work. Then, open your book and try again. You get **5,000** points for each additional correct answer. So, **show what you know!**

1. Name the main parts of the hydrologic cycle.

 ______________________________________________

 ______________________________________________

2. What is the name of the process by which water soaks into the ground and becomes groundwater?

 ______________________________________________

3. What is the name of the half of our Earth that is located above the equator?

 ______________________________________________

4. What is the name of the half of our Earth that is located below the equator?

 ______________________________________________

5. The warm waters around the equator are called the ____________________

6. There are two main types of water on Earth. Oceans and seas are ______________,

 and rivers, streams, and most lakes are ______________________________.

7. Ground is said to be ______________ when it is holding as much water as it can.

8. The largest amount of freshwater on our Earth is found in what frozen formations?

_______________________________________________

9. A layer of hard-packed snow and ice on a glacier is called a ____________________.

10. An ________________ is a large chunk of ice that breaks off of a glacier.

11. The process by which the above happens is called ____________________________.

First Attempt ________________
*(number of correct responses X 10,000)*

Second Attempt + ________________
*(number of correct responses X 5,000)*

**Total Number of Points** = ________________

Rain
Fair

# Unit Four
# The Atmosphere

## Unit Timeline

1593 Galileo Galilei invents the thermometer.

1640s Evangelista Torricelli discovers air pressure and invents the barometer.

1657 The mercury thermometer is invented.

Mid-1600s Blaise Pascal discovers that air pressure decreases with altitude.

1749 Alexander Wilson attempts to measure upper air temperature using a kite and a thermometer.

1783 The Montgolfier brothers invent the first hot air balloon and conduct the first flights.

1783 John Jeffries studies air at various altitudes with balloon flights and a barometer.

1804 Joseph Louis Gay-Lussac measures air at 23,000 feet.

1902 Teisserenc de Bort measures air at 40,000 feet and records temperatures of –67 degrees Fahrenheit. He names the troposphere and stratosphere.

## Unit Four Vocabulary

- air composition
- atmosphere
- atmospheric pressure
- barometer
- composition
- dehydrated
- firmament
- hetero
- heterosphere
- homos
- homosphere
- humidity
- hygrometer
- mercury
- mesosphere
- psychrometer
- relative humidity
- stratosphere
- temperature
- thermosphere
- Torricelli, Evangelista
- troposphere

## Materials Needed for This Unit

- Science notebook
- 2 photocopies of the "Checking It Out" form in Appendix A
- Photocopy of the "Observation Form" in Lesson 13
- Photocopy of the "Form for Recording Temperatures" in Lesson 14
- 8 to 10 sheets of 8½-by-11-inch blue construction paper (I recommend using several different shades for appearance.)
- Index cards
- Markers and crayons
- "Sticky tack" removable adhesive
- Glass with straight sides
- 12-inch plastic ruler
- Strong tape that can withstand water
- Clear straw
- Stick of chewing gum
- Food coloring
- 2 centigrade, or Celsius, thermometers
- Tape
- Wet gauze or cotton ball
- Thin rubber band
- Fan (You can use a nearby ceiling fan or even wave magazines to simulate a fan.)
- Tape measure
- Colored candy (optional)
- Paper plate — 1 per student
- Encyclopedia and dictionary
- Materials for Unit Wrap-Up (See page 151.)

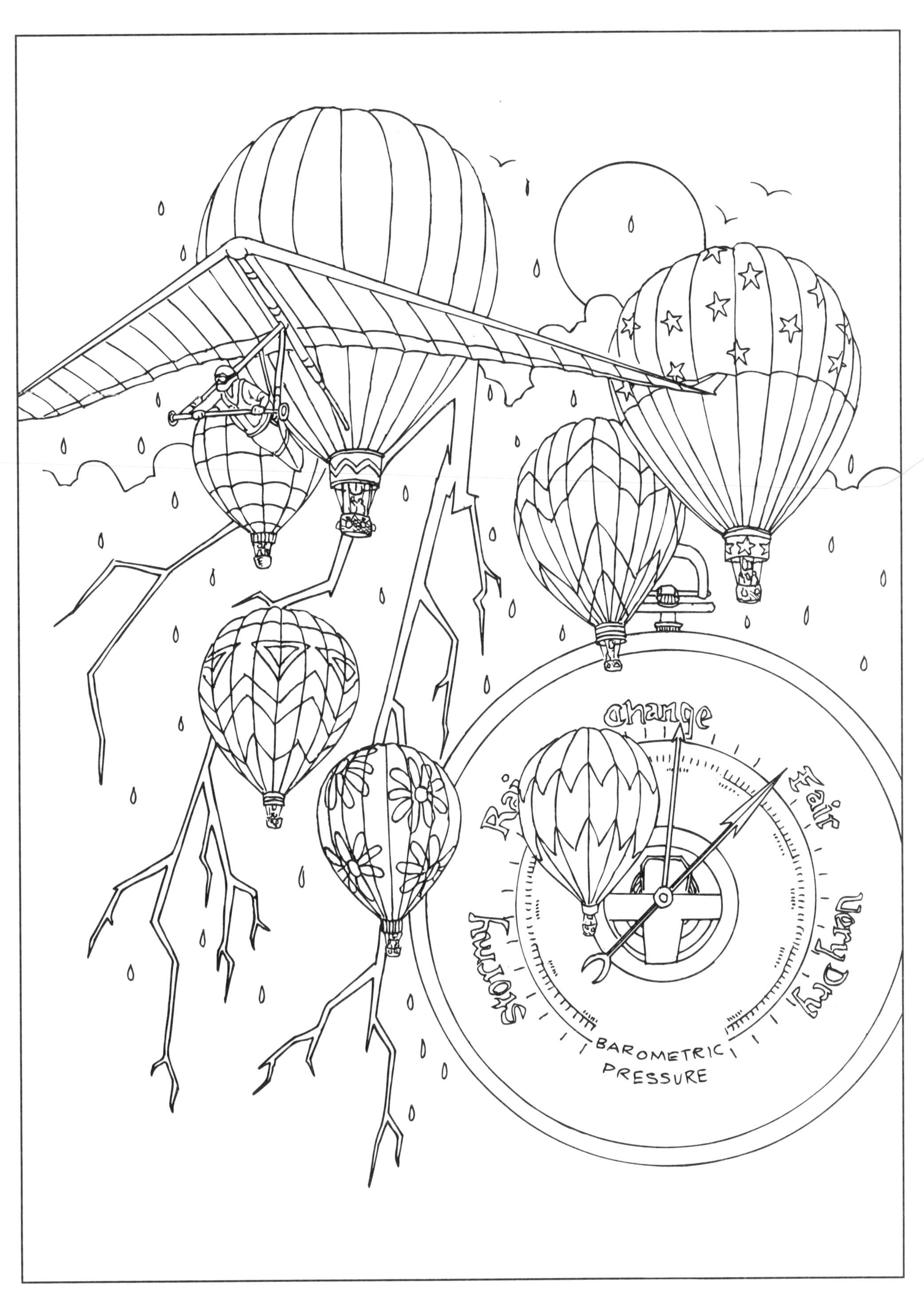
change
Rai
Fair
Very Dry
Stormy
BAROMETRIC
PRESSURE

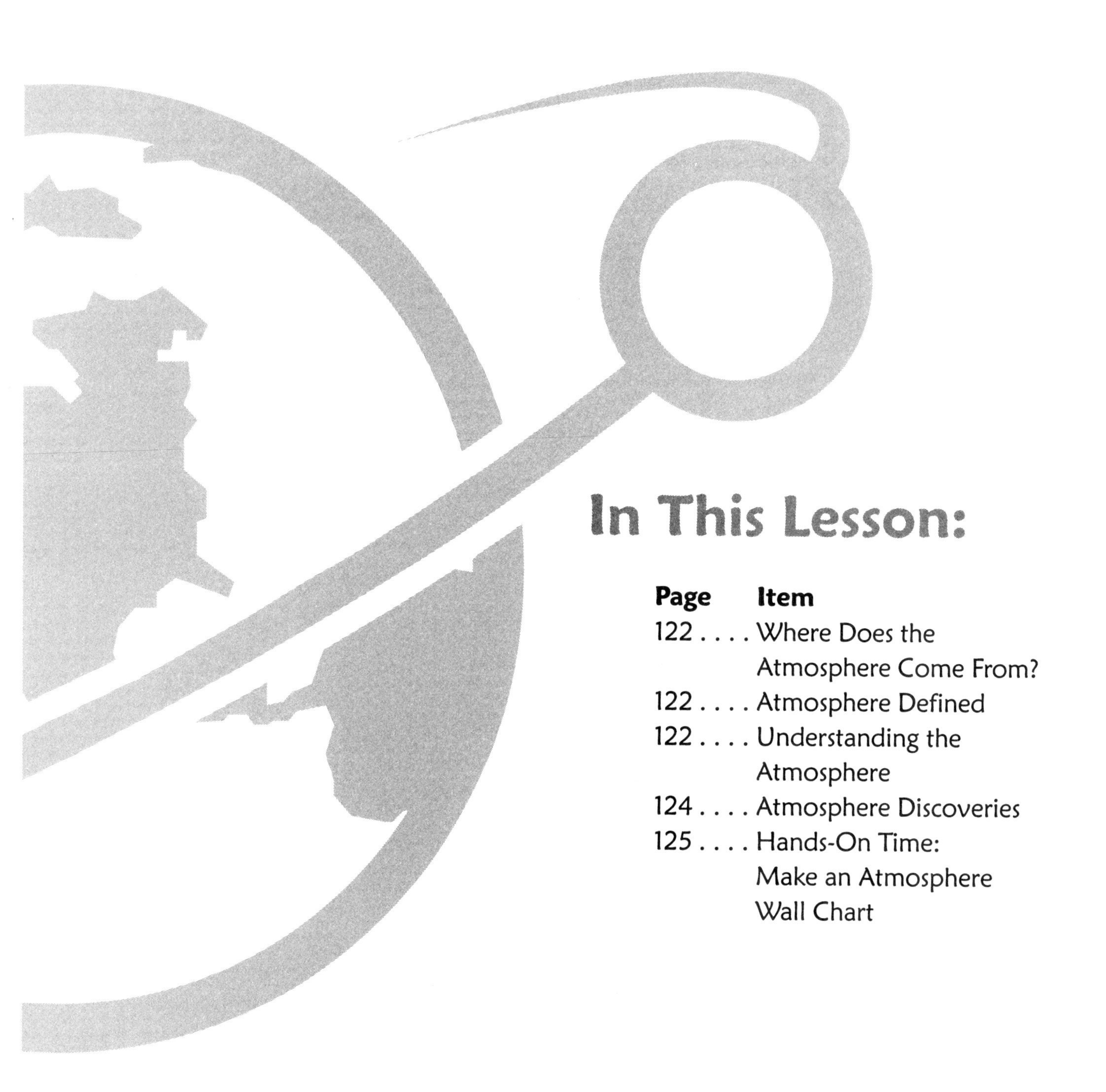

# In This Lesson:

# Lesson 12

# INTRODUCTION

## Teaching Time:

## Layer by Layer

This week we are beginning a brand new unit. So far in this book, we've learned some basics about Earth; we've studied the lithosphere; and we've studied the hydrosphere. Each of these topics has presented us with information about the world God created for us. This new unit takes us on an airy adventure. We're going to learn about the atmosphere. There is a lot to learn, and I think we're going to have a lot of fun. I hope that you are able to see God's design in each aspect of our Earth. He is an orderly and reasonable God. He lovingly provided us with many ways to learn more about Him and His love for us. We really can trust Him with everything!

*Note to Teacher*: In the next unit, advanced students will use instruments made in the Hands-On Time activities in this unit. So, please do not discard these instruments!

"Praise Him, you heavens of heavens, and you waters above the heavens! Let them praise the name of the LORD, for He commanded and they were created." (Psalm 148:4–5)

**firmament**
An expanse; the sky and everything beyond Earth, including the Sun, Moon, and stars.

**atmosphere**
The gases, necessary to support life, that surround our planet.

**composition**
One of two methods of studying the atmosphere; breaks the atmosphere into two layers, the homosphere and the heterosphere.

**homosphere**
One of two layers that make up the composition of the atmosphere; composed of the troposphere, stratosphere. and mesosphere.

**heterosphere**
One of two layers that make up the composition of the atmosphere.

**temperature**
One of two methods of studying the atmosphere; breaks the atmosphere into four layers, the troposphere, stratosphere, mesosphere, and thermosphere.

How hot or cold the air is.

## Where Does the Atmosphere Come From?

The Bible tells us that on the second day of creation "God said, 'Let there be a firmament in the midst of the waters, and let it divide the waters from the waters.' " (Genesis 1:6) Genesis 1:8 says, "And God called the firmament Heaven." I need you to put on your thinking caps here. I'm going to teach you a little something about understanding the Bible. I did some checking using some trustworthy Bible study materials, and I learned that another word for **firmament** is *expanse*. This is referring to the area above Earth. In other words, it is actually referring to the sky.

## Atmosphere Defined

Part of this firmament is our **atmosphere**. The atmosphere consists of the gases that are around our planet. These gases are absolutely necessary for life on Earth. The atmosphere provides the oxygen we need to breathe. It helps keep temperatures in the right range on Earth. It even protects us from meteors! I'd say our atmosphere is very important.

## Understanding the Atmosphere

Scientists study the atmosphere in two different ways. The first way is by looking at the makeup, or **composition**, of the atmosphere. The composition method divides the atmosphere into two major layers. These layers are called the **homosphere** (**home** us fear) and the **heterosphere** (**het** er us fear).

The other way to study the atmosphere is by studying the **temperature** of the atmosphere at different heights. In the temperature method, the atmosphere is divided into four layers. These layers are called the troposphere, the stratosphere, the mesosphere, and the thermosphere. The troposphere, stratosphere, and mesosphere combine to form the homosphere.

These are the layers closest to the surface of Earth. Keep reading to learn more.

## The Troposphere

The **troposphere** (**trop** us fear) is the lowest layer of the atmosphere. It goes up to almost 7 miles above Earth. The temperature goes down at a steady rate the higher one travels in this layer. At the highest point of the layer, the temperature is about –50 degrees Celsius. This is the layer that produces our weather, even our storms. You will learn more about storms in another unit.

## The Stratosphere

The **stratosphere** (**strat** us fear) is the next layer of the atmosphere. It goes from about 7 miles above Earth to approximately 31 miles above Earth. The temperatures in this layer actually increase as one travels higher. The temperature range is –50 degrees Celsius to 0 degrees Celsius. The stratosphere is dust-free and cloud-free. It is much calmer than the troposphere.

## The Mesosphere

The **mesosphere** (**mez** us fear) begins at 31 miles above Earth and ends at about 50 miles above Earth. Winds in this layer can be very strong. Temperatures in the mesosphere go down again, with the range being 0 degrees Celsius down to -78 degrees Celsius.

## The Thermosphere

The last layer of the atmosphere is called the **thermosphere** (**therm** us fear). This layer begins at the end of the mesosphere. In the thermosphere, temperatures increase at a rapid rate the higher one travels since the Sun is ever nearer. It is quite warm during the daytime but gets very cold at night. In the thermosphere, the meteoroids that could endanger us vaporize, forming what we call "shooting stars."

### Name It!

**troposphere**
Part of the temperature method of studying the atmosphere, this is the lowest layer, extending almost seven miles above Earth.

**stratosphere**
Part of the temperature method of studying the atmosphere, this layer extends from about 7 miles to about 31 miles above Earth. It is just above the troposphere.

**mesosphere**
Part of the temperature method of classifying the atmosphere, this layer is from 31 to 50 miles above Earth.

**thermosphere**
Part of the temperature method of studying the atmosphere, this layer begins after the mesosphere.

Additional Notes

## Atmosphere Discoveries

Of course, there is much more to learn about the atmosphere. Some of it will be covered in the remaining lessons of this unit. Some you will learn as you get older, and some you will have to discover on your own. You can look at the two sections that follow to see some of the discoveries regarding our atmosphere. As with almost everything in this book, there is much more to learn than one person can teach you. You get the fun of seeing how much more you can learn on your own! You might want to give thanks to God for the beautiful, detailed world He created for you and for me.

### Atmosphere Is Discovered

- **1593**: Galileo Galilei (1564–1642) invented the thermometer.
- **1640s**: Evangelista Torricelli discovered air pressure and invented the barometer.
- **1657**: Mercury thermometer was invented.
- **Mid-1600s**: Blaise Pascal discovered that air pressure decreases with altitude.

### Air Temperature Is Measured

- **1749**: Alexander Wilson (1714–1786; Scotland) attempted to measure upper air temperature using a kite and a thermometer.
- **1783**: Montgolfier brothers (France) invented the first hot air balloon and conducted the first flights.
- **1783**: John Jeffries (1745–1819; America) studied air at various altitudes with balloon flights and a barometer.
- **1804**: Joseph Louis Gay-Lussac (1778–1850) measured air at 23,000 feet.
- **1902**: Teisserenc de Bort measured air at 40,000 feet and recorded temperatures of -67 degrees Fahrenheit; named the troposphere and stratosphere.

Additional Notes

## Hands-On Time:

# Make an Atmosphere Wall Chart

## Featured Activity

### Objective

To create a tool to help learn the characteristics of the atmospheric layers.

### Materials

- 8 to 10 sheets of 8½-by-11-inch blue construction paper (I recommend using several different shades for appearance.)
- Index cards
- Markers/crayons
- "Sticky tack" removable adhesive

### Instructions

1. Using sticky tack, adhere paper to wall beginning at the baseboard. You'll need to apply the paper end-to-end working your way up the wall as high as you can go.

2. Write the name of each layer on one card, in large letters. (Troposphere, Stratosphere, Mesosphere, Thermosphere)

3. For each of these cards, make another card that lists the traits of that layer.

**Additional Notes**

4. With sticky tack on the back of each card, you can practice putting your cards in the appropriate order on the "chart." Have your teacher help you put the thermosphere cards up high.

5. This will help you memorize the layers of the atmosphere and their traits.

6. Have fun!

For more fun: Decorate your "sky." Add drawings of airplanes, birds, clouds, etc. Also, near the floor, you can add grass, trees, and buildings.

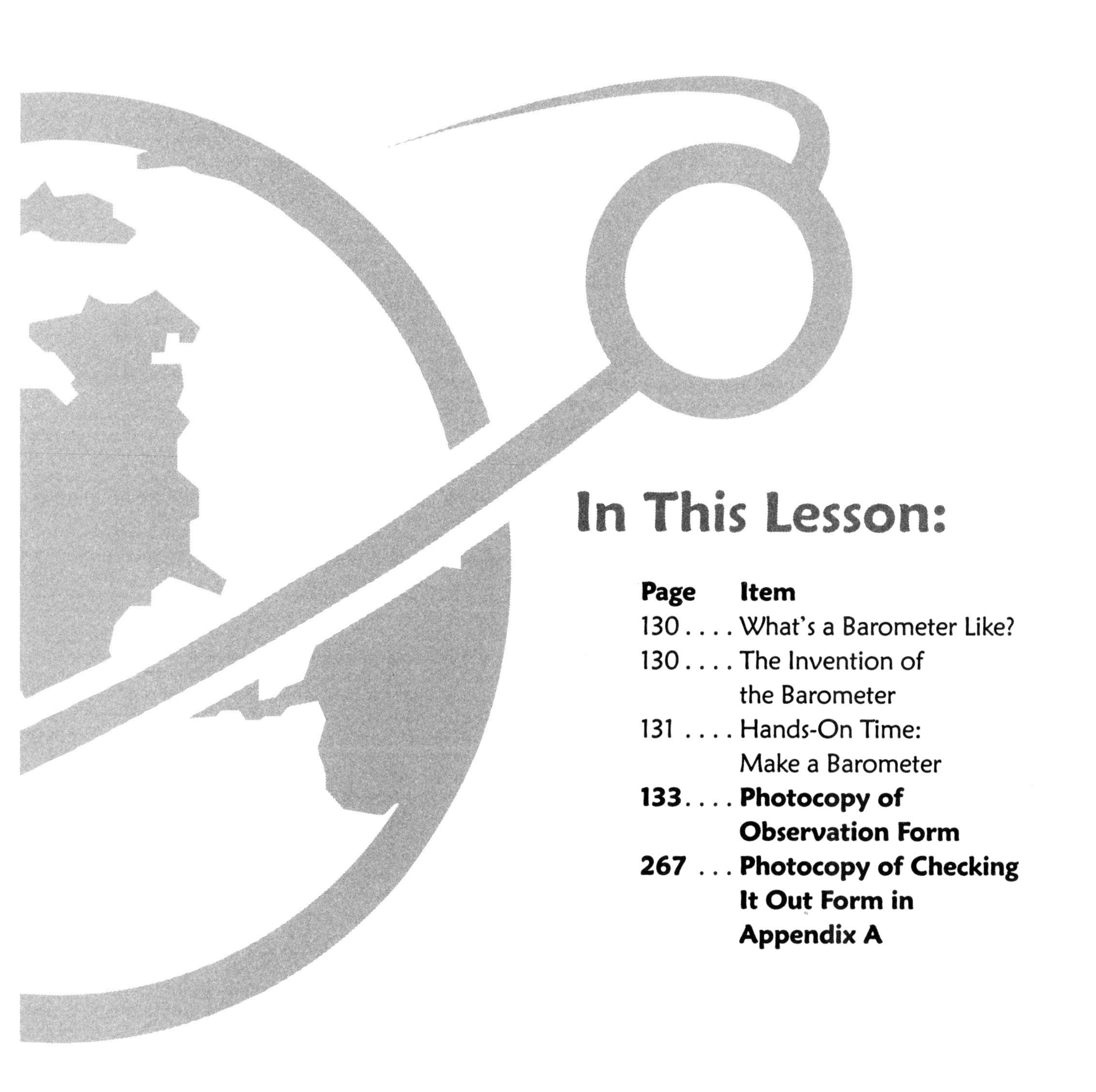

# In This Lesson:

Lesson 13

# ATMOSPHERIC PRESSURE

## Teaching Time:
# Feel the Pressure

Would you believe me if I told you that right now you're being squeezed from every direction? Well, you are. I suppose you're wondering exactly what is pressing on you. The fact is that air is constantly pressing on us all. We don't feel like we're being squeezed because it's happening to all sides of our body evenly at the same time. This is called **atmospheric pressure**, and it's one of the many interesting things about the air around us. We're going to learn about it today as we continue our study of the atmosphere.

Atmosphere and weather are closely related. Therefore, we're going to begin by talking about weather to help us understand atmospheric pressure. Do you remember which layer of the atmosphere our weather takes place in? It's the troposphere. A nickname for the troposphere is the "weather layer." In the troposphere, there is a shifting of warm air and cold air. Warm air is lighter than cool air and it rises. Cold air, being heavier than warm air, sinks. As more cold air moves into a region, the

"Let them praise the name of the LORD, for His name alone is exalted; His glory is above the earth and heaven." (Psalm 148:13)

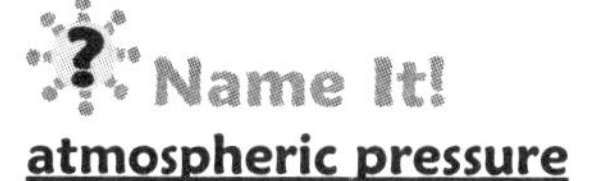

**atmospheric pressure**
The pressure of the air all around us.

**mercury**
A heavy toxic liquid.

**barometer**
Instrument that measures changes in air pressure.

**Evangelista Torricelli**
Inventor of the barometer.

cold air sinks, causing an increase in air pressure, or atmospheric pressure. (You'll want to remember this information for our weather unit.) Using a **barometer** (buh **rom** eh ter), an instrument that measures changes in air pressure, weather forecasters can predict upcoming changes in the weather.

## What's a Barometer Like?

I don't know if you caught that, so I'll repeat it. Changes in air pressure can be an indication of changing weather! This excites me. It makes me realize how great God is. How does this relate to barometers? A barometer is a glass tube filled with **mercury**, a heavy, toxic liquid that has also been used in thermometers. (Its use in thermometers has decreased in recent years.) The tube is closed on one end and open on the other. The open end is placed in a container that also has mercury in it. When the air pressure changes, so does the level of the mercury in the tube. There is more to the story, but this is the basic information. These changes tell the scientists who study them what is going on in the atmosphere regarding pressure. This information helps them then forecast the weather.

## The Invention of the Barometer

Since we've learned that barometers can help us predict weather changes, let's learn who invented the barometer. **Evangelista Torricelli** is the inventor. He lived during the 1600s in Italy and knew a famous scientist known to us as Galileo. As a matter of fact, Torricelli was Galileo's assistant. In the 1640s, following Galileo's advice, Torricelli filled a four-foot-long tube with a heavy liquid called mercury. He turned the tube upside down on a dish. He studied the results. One of the results he noticed was changes in the height of the liquid as the air pressure changed. This is known as the first barometer. You will make a very simple barometer in your Hands-On Time activity.

There's much more to atmospheric pressure and barometers

than we can study at this level. Don't you find it interesting to know how much we can learn about our world? I hope that you get excited every time you discover there is more to God's creation than you realized. Up until this unit, when you were outside, you may have looked up and thought there was just a sky. You've learned that there is an atmosphere, made up of a homosphere and a heterosphere. The homosphere has a troposphere, a stratosphere, and a mesosphere. The troposphere has weather. The atmosphere exerts pressure on our Earth and all who live here. And you thought there was just a sky! Just wait. As always, there's more to learn another day. God is indeed good!

### Discovery Zone

See how barometric pressure changes in the life of a storm (and more!) at the "Exploring Earth" Web site:

www.classzone.com/books/earth_science/terc

Choose: Select a Chapter/Chapter 19/Visualizations/Examine how barometric pressure changes with weather conditions.

Or, you can enter the keycode ES1902 where directed on the opening screen.

## Hands-On Time:
# Make a Barometer

## Featured Activity

### Objective

To observe the changes in pressure by watching the changes in our liquid.

### Materials

- Glass with straight sides
- 12-inch plastic ruler
- Tape (You will need a strong tape that can withstand water.)
- Clear straw

- Stick of chewing gum
- Water
- Food coloring
- Photocopy of the "Checking It Out" form in Appendix A
- Photocopy of the "Observation Form" in this lesson

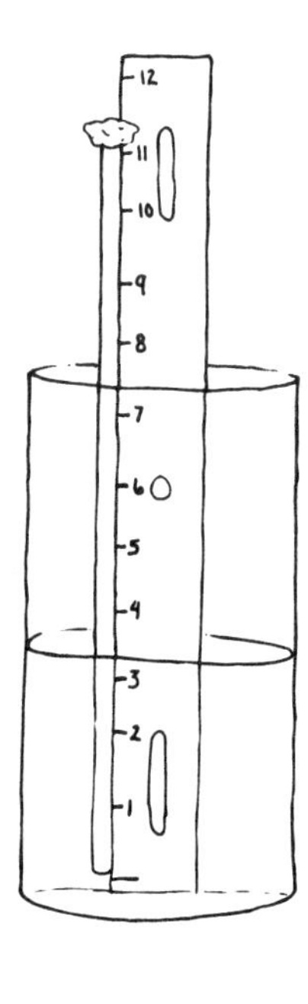


## Instructions

(*See the illustration accompanying this activity.*)

1. Stand the ruler inside the glass with the markings pressed against the glass, so you can read them through the glass. Tape in place.

2. Place the straw inside the glass *without* letting it touch the bottom of the glass and put it up against the ruler. Tape it to the ruler.

3. Begin chewing the gum. Really. If you can't chew gum, ask your teacher for help with this part.

4. Fill the glass about half full of water. (Be sure the bottom of the straw is in the water.) Add a few drops of food coloring.

5. Suck on the straw to bring water up about halfway into the tube. Stop up the straw with your finger and then get your gum to plug the tube. You want water to remain in the tube.

6. In the "Observation Form" provided, record the measurement of the water in the tube.

7. Complete a "Checking It Out" form and place it in your science notebook.

8. Take a reading each day and record the measurement.

9. Also, note the weather for each day.

10. Can you draw any conclusions based on the weather and the level of water in the straw?

11. Repeat this over the course of one week. Do you notice any relation between the water level and the weather?

12. Complete the "Observation Form" and place it in your science notebook with your "Checking It Out" form for this experiment.

## Additional Notes

**Observation Form**

| | Mon | Tues | Wed | Thurs | Fri | Sat | Sun |
|---|---|---|---|---|---|---|---|
| Date | 1/22 | | | | | | |
| Water Level Measurement | 7 1/4" | 7 1/4" + 1/16 | | | | | |
| Weather Conditions (sunshine, clouds, rain, snow, storms) | clear sunshine warm | clear skies, sunny cooler | | | | | |

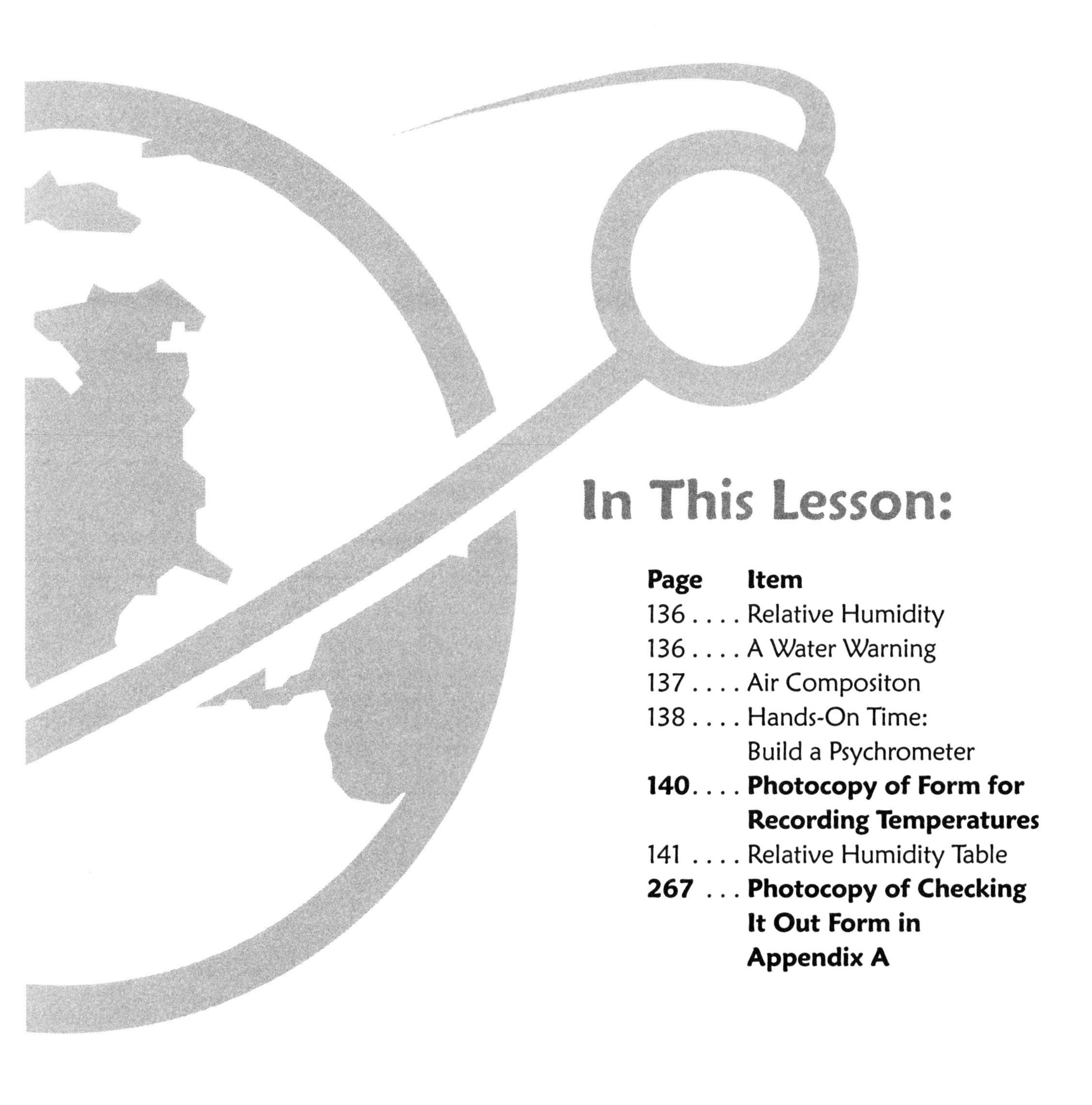

# In This Lesson:

Lesson 14

# HUMIDITY

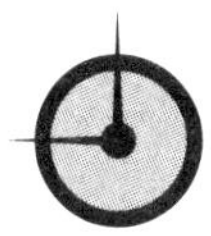

## Teaching Time:

## Moisture in the Air

Today we are going to continue our study on the atmosphere. You've had the opportunity, so far, to learn about the layers of the atmosphere and atmospheric pressure. Hopefully, you're getting the picture that there is much more to the air that surrounds Earth than meets the eye. After all, you really cannot see the air. It's there and we know that. We breathe it every day. We must have it to live. However, it could be easy to miss all the marvelous details God created when He created this wonderful thing called the atmosphere . . . unless you study it.

The question I want you to consider today is: What does the air around us consist of? Have you ever thought of that before? Can you make an educated guess about what makes up our atmosphere? Now would be a good time to discuss this with your teacher.

The topic we're going to study today is **humidity** (hue **mid** i tee). Have you ever heard that word before? Perhaps you've heard people discuss how humid it is outside. Often people are

**Scripture**

"...whereas you do not know what will happen tomorrow. For what is your life? It is even a vapor that appears for a little time and then vanishes away." (James 4:14)

**Name It!**

**humidity**

Moisture content of the air.

**relative humidity**
The amount of moisture the air can hold at a specific temperature.

complaining that the humidity is too high, and they feel hot or uncomfortable because of it. In some parts of the southern United States, some people say, "It's close." They mean it's humid. Humidity is the moisture content in the air. Air actually holds moisture. It is in the form of water vapor, so you should not expect to notice it.

The more moisture that is in the air, the more humid it is. When it is very humid, we usually feel hotter than when the humidity is low. Isn't that interesting? The reason for this is that God has created our bodies with an ability to cool us by the sweating process. When the humidity is high, there is already a lot of moisture in the air. This means that our sweat will not evaporate quickly. Instead, it stays on our skin. This makes us feel hot. When the humidity is low, the sweat evaporates much more quickly, making us feel cooler. (The evaporation process actually uses up heat molecules.) Now this may seem complicated to you. Mainly, you can remember that when sweat lingers on your skin and your clothes feel damp, it's a sign that the humidity is high.

## Relative Humidity

It's important to know that weather forecasters and weather reports usually refer to **relative humidity**. Relative humidity is the amount of moisture the air can hold at a specific temperature. In other words, there is only so much moisture the air can hold at a time. At a certain point the air becomes saturated and must release some of the moisture. This can happen in the form of rain, mist, or snow, to name a few. When the air is saturated, the relative humidity is 100 percent.

## A Water Warning

So far we've talked more about high humidity than low humidity. Let's focus on low humidity for a moment. When the humidity is low, you will usually feel cooler than you would if the temperature was the same, but the humidity was high. As I mentioned, this is because the sweat evaporates, cooling us down. Now when this happens, you might be fooled into thinking you're not sweating much, even if the temperature is very high. This could cause you to get **dehydrated** (dee **high** dray ted). This is a serious medical condition in which your body does not have enough water to function properly. I'm telling you this, not to frighten you, but to remind you that it is important to drink plenty of water, especially when temperatures are high. This is true even if you do not feel hot. Water is essential to good health, and your body requires a lot of it. Sodas and juices cannot replace the benefits of water, so drink up!

**Name It!**

**dehydrated**
Serious medical condition in which your body does not have enough water to function properly.

**air composition**
The "ingredient list" of the air.

## Air Composition

I have told you that water vapor is always present in the air. However, it is not listed when giving the **air composition**, or "ingredient list," for Earth's air. The reason is that the amount of water vapor is always changing. Therefore, the air composition is listed apart from water vapor. You will learn more about this in your next lesson.

Even though we covered some technical things today, I hope you understood the basics. Air is a real substance that is made up of real things. One of the things that is always present in the air is moisture. We call the moisture in the air the humidity. God is detailed even in the creation of air.

If you don't have centigrade thermometers, you can convert the temperature:

Fahrenheit =
Celsius x 9/5 + 32

Celsius =
(Fahrenheit – 32) x 5/9

**hygrometer**
Instrument used to measure humidity.

**psychrometer**
A type of hygrometer that measures the relative humidity.

# Hands-On Time: Build a Psychrometer

## Featured Activity

### Objective

To learn how scientists discover the relative humidity each day.

### Materials

- 2 centigrade, or Celsius, thermometers
- Tape
- Wet gauze or cotton ball
- Thin rubber band
- Fan (You can use a nearby ceiling fan or even wave magazines over the thermometers to simulate a fan.)
- Photocopy of the "Form for Recording Temperatures" in this lesson
- Photocopy of the "Checking It Out" form in Appendix A

### Instructions

Humidity is measured by a type of instrument called a **hygrometer** (high **grom** eh ter). There are a few types of hygrometers. One type is called a **psychrometer** (sigh **krom** eh ter), which gives the relative humidity. This type takes two temperatures, one with a wet thermometer and one with a dry thermometer. Room temperature air is blown over both thermometers and each is read. When the reading on the wet thermometer is subtracted from the reading on the dry one, you can use a table and determine if the humidity is high or low. Today you get to make a psychrometer of your own and evaluate the humidity where you are.

1. Tape both thermometers to a flat surface near your fan. The kitchen counter would work well.

2. Using the rubber band, secure the wet cotton onto the bulb of one of the thermometers.

3. Turn on the fan and blow on the thermometers until the temperatures are reading steady.

4. Read each thermometer and record the readings on the "Form for Recording Temperatures" provided here. Keep the form so that you can repeat this for several days; try moving to different locations to see the differences.

5. Subtract the wet temperature from the dry temperature.

6. Using the "Relative Humidity Table" in this lesson, find the relative humidity and list it on your form. The column on the left is the dry bulb reading. The row across the top is the difference between the two readings. To find the relative humidity, locate your dry bulb reading in the left column. Go across the table until you're just under the dry-bulb-minus-wet-bulb reading. This is the relative humidity. As you can see, the greater the difference is, the lower the relative humidity is.

7. Complete a "Checking It Out" form and place it in your science notebook along with your recorded findings.

**Additional Notes**

## Form for Recording Temperatures

| Date | Location | Dry Bulb Temp °C | Wet Bulb Temp °C | Dry minus Wet | Relative Humidity % |
|---|---|---|---|---|---|
| | | | | | |
| | | | | | |
| | | | | | |
| | | | | | |
| | | | | | |

## Relative Humidity Table (in percent)

| Dry Bulb | Dry Bulb Minus Wet Bulb (degrees Celsius) | | | | | | | | |
|---|---|---|---|---|---|---|---|---|---|
| **°C** | **1** | **2** | **3** | **4** | **5** | **6** | **7** | **8** | **9** | **10** |
| **10** | 88 | 77 | 66 | 55 | 44 | 34 | 24 | 15 | 6 | - |
| **11** | 89 | 78 | 67 | 56 | 46 | 36 | 27 | 18 | 9 | - |
| **12** | 89 | 78 | 68 | 58 | 48 | 39 | 29 | 21 | 12 | - |
| **13** | 89 | 79 | 69 | 59 | 50 | 41 | 32 | 22 | 15 | 7 |
| **14** | 90 | 79 | 70 | 60 | 51 | 42 | 34 | 25 | 18 | 10 |
| **15** | 90 | 81 | 71 | 61 | 53 | 44 | 36 | 27 | 20 | 13 |
| **16** | 90 | 81 | 71 | 63 | 54 | 46 | 38 | 30 | 23 | 15 |
| **17** | 90 | 81 | 72 | 64 | 55 | 47 | 40 | 32 | 25 | 18 |
| **18** | 91 | 82 | 73 | 65 | 57 | 49 | 41 | 34 | 27 | 20 |
| **19** | 91 | 82 | 74 | 65 | 58 | 50 | 43 | 36 | 29 | 22 |
| **20** | 91 | 83 | 74 | 67 | 59 | 53 | 46 | 39 | 32 | 26 |
| **21** | 91 | 83 | 75 | 67 | 60 | 53 | 46 | 39 | 32 | 26 |
| **22** | 91 | 83 | 76 | 68 | 61 | 54 | 47 | 40 | 34 | 28 |
| **23** | 92 | 84 | 76 | 69 | 62 | 55 | 48 | 42 | 36 | 30 |
| **24** | 92 | 84 | 77 | 69 | 62 | 56 | 49 | 43 | 37 | 31 |
| **25** | 92 | 84 | 77 | 70 | 63 | 57 | 50 | 44 | 39 | 33 |
| Suitable for up to 77°F | | | | | | | | | | |

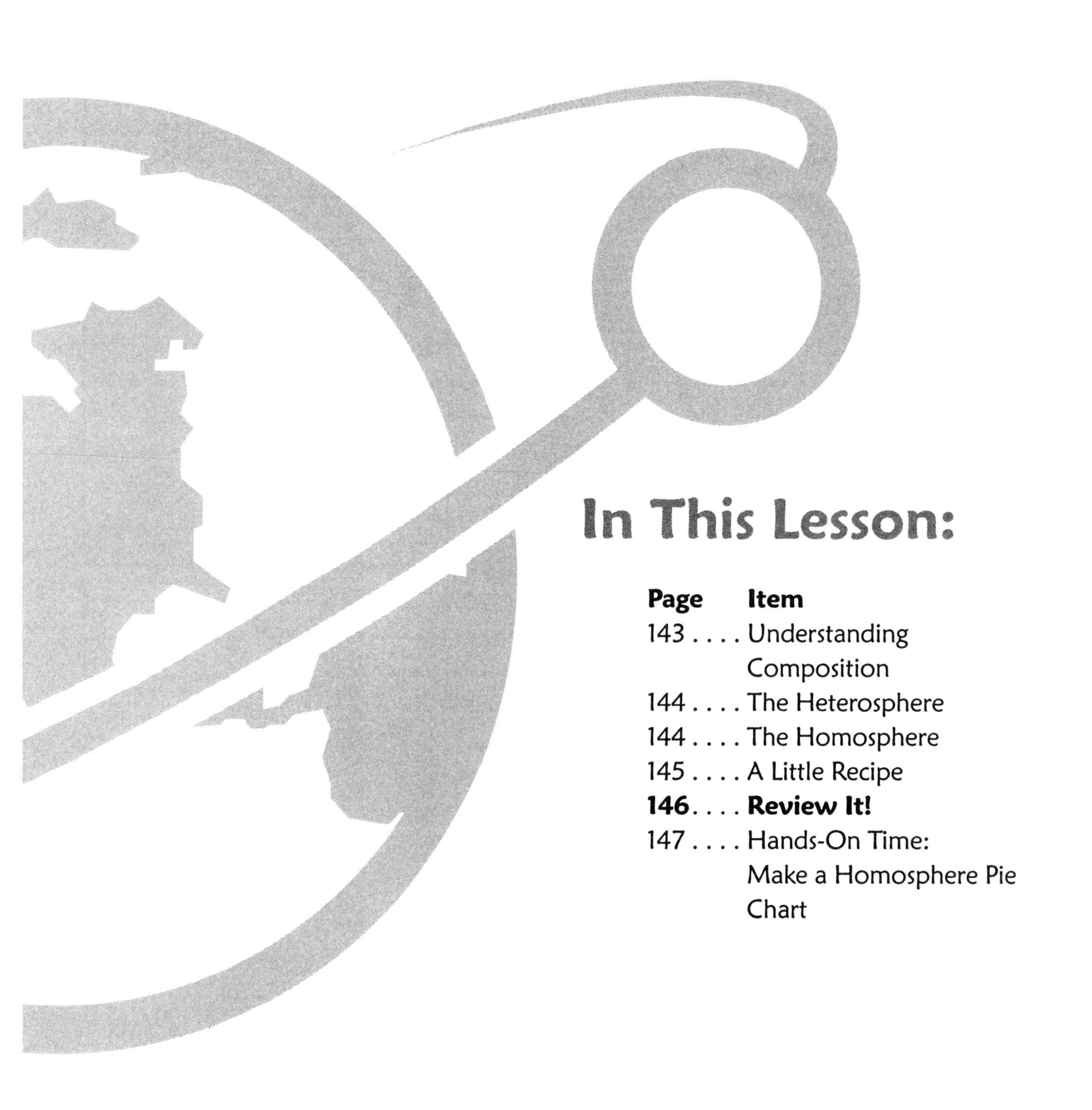

# In This Lesson:

| Page | Item |
|---|---|
| 143 . . . . | Understanding Composition |
| 144 . . . . | The Heterosphere |
| 144 . . . . | The Homosphere |
| 145 . . . . | A Little Recipe |
| **146** . . . . | **Review It!** |
| 147 . . . . | Hands-On Time: Make a Homosphere Pie Chart |

Lesson 15

# AIR COMPOSITION

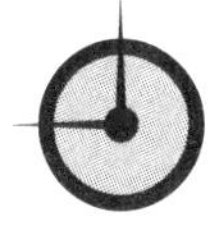

Teaching Time:

## The Air We Breathe

I'm curious. I'm wondering just how much you know about the air around you. Of course, I know what we've studied in this unit so far. I know that you have learned that air is more complex than you may have thought up until now. For instance, you've learned that air has moisture in it. That's called humidity. You've learned that air makes up our atmosphere and that the atmosphere can be divided into layers according to temperature and also according to composition. What does composition mean? That's what we're going to learn about today.

**Scripture**
"And the LORD God formed man of the dust of the ground, and breathed into his nostrils the breath of life; and man became a living being." (Genesis 2:7)

### Understanding Composition

Using the composition method, I told you that the atmosphere could be divided into two layers, the homosphere and the heterosphere. As you may recall, that is all we discussed on that subject. Let's learn more. In Lesson 12, you learned that the atmosphere is the gases that surround our planet. First of all,

**hetero**
Different.

**homos**
Same.

these gases are hydrogen, helium, oxygen, and nitrogen, as well as some others in very small amounts. Second, the word *composition* is from a Latin root meaning "a putting together." In other words, it is like a recipe. The homosphere and the heterosphere have different compositions, or recipes, based on what gases are found in each layer.

## The Heterosphere

The root word **hetero** means "different." When referring to the atmosphere, it means that the composition is different according to the height in the layer. There are different sections of the heterosphere. Lower parts of the heterosphere have more oxygen in their composition. The composition of the next higher area has more helium than other gases, and the highest part has more hydrogen. You might want to concentrate on remembering that heterosphere means different compositions. One more thing, where does the heterosphere begin? It begins at about 50 miles above Earth. That means there is something below. Let's go exploring!

## The Homosphere

Aha! I can imagine you knew this was coming. After all, you already know that there are only two layers of the atmosphere according to the composition method. Of course, the homosphere is the part of the atmosphere below 50 miles. Let's look at the root word for this layer. The root word **homos** means "same." Since we just studied the heterosphere and learned that this means the composition of the air is different in different parts, it makes sense to assume that the homosphere is basically the same in composition throughout. And that *is* a safe assumption. The composition of the homosphere is consistent throughout.

Before we go further, I want to remind you that you have already learned a lot about the homosphere. The homosphere is made up of most of the temperature layers that we have already studied. The troposphere, the stratosphere, and the mesosphere are all parts of the homosphere. Recall that you studied these layers of the homosphere in Lesson 12. You might want to refresh your memory by reading that lesson again at this point.

## A Little Recipe

The composition of the homosphere is the last thing we need to look at today. The homosphere is the air that you and I breathe. In health classes, you may have learned that humans need oxygen to breathe. Therefore, you would be correct in guessing that the homosphere has oxygen in it. What might surprise you is how much (or how little) of the homosphere is oxygen. Only 21 percent of the homosphere is oxygen. The "Composition of the Homosphere" pie chart in this lesson reveals the rest of the story. What do you see when you look at the pie chart? What is the homosphere *mostly* composed of? Yes, nitrogen. Nitrogen is important in making fertilizer, explosives, medicines, and plastics. The tiniest portion of the pie chart shows argon and other gases as the remaining part of the homosphere's composition, or recipe. You might be tempted to think that these aren't important, but do you know that argon is the gas used in light bulbs? I don't know about you, but I think light bulbs are wonderful and very important in our world.

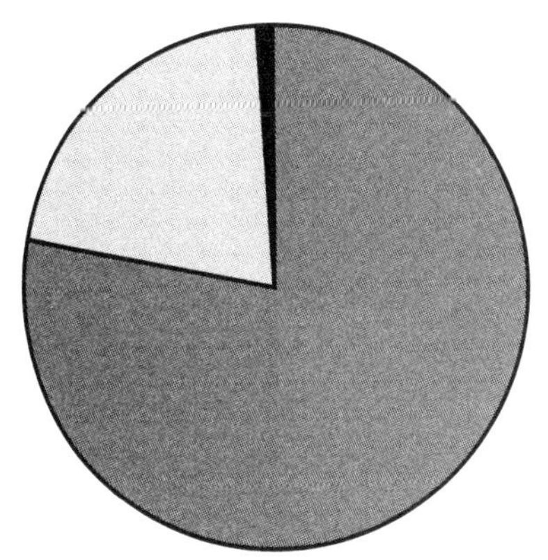

**Composition of the Homosphere**

**Legend**
■ Nitrogen (78%)
□ Oxygen (21%)
■ Other, including argon (1%)

Additional Notes

## Review It!

*May be done orally or written.*

1. What are the two ways the atmosphere can be grouped?

    ______________________________

2. What does composition mean?

    ______________________________

    ______________________________

3. What are the two layers using the composition method?

    ______________________________

    ______________________________

4. What does the root for each word mean?

    ______________________________means

    ______________________________

    ______________________________means

    ______________________________

5. Which layer is the air we breathe?

    ______________________________

6. How much of the air we breathe is oxygen?

    ______________________________

Additional Notes

## Hands-On Time:

# Make a Homosphere Pie Chart

## Featured Activity

### Objective

To review the composition of the air we breathe.

### Materials

- Tape measure
- Paper plate — 1 per student
- Markers or crayons
- Pencil
- Colored candy (optional)

### Instructions

Today we're going to duplicate the pie chart showing the composition of the homosphere.

1. Measure around the plate using a tape measure. Mine is 28 inches in circumference; that is, 28 inches around the plate.

2. Make a mark in the center of the plate.

3. Oxygen is about one-fifth of the homosphere's composition. Therefore, 28 inches divided by 5 is about 6½ inches. Make marks in two places that are about 6½ inches apart.

**Helpful Hint**
Remember, the pie chart represents all (100%) of the air around us. The other percentages represent portions of all the air.

4. At these points, draw lines to the center of the plate. Color this wedge and label it "Oxygen — 21%."

5. At the edge of the plate, measure out another ½ inch. Draw another line to the center. Color this section another color and label it "Argon and other gases — 1%."

6. Color the remaining large portion of the plate a third color and label it "Nitrogen — 78%."

7. For a variation: Instead of coloring the plate, fill each section with colored candy. For example, all red in one section, all green in another, and so on.

### Another Option

Draw the pie chart on a piece of paper, coloring in the sections as noted. Label the diagram "Composition of the Homosphere." Add this drawing to your science notebook.

## Activity for Advanced Students: Discover More!

Choose at least one of the mentioned gases to research. Using an encyclopedia, the Internet, and a dictionary, discover what you can about this gas. Write a paragraph about what you learned. Add this to your science notebook.

# Unit Four Wrap-Up

Now that you've completed the unit, it's time to wrap up all that you've learned and show what you know!

1. Copy this page and place a check mark in each box for work you've completed.
   - ❑ Lessons read
   - ❑ Daily Reading Sheets
   - ❑ Hands-On Time
   - ❑ Checking It Out forms
   - ❑ Vocabulary
   - ❑ Coloring Page

2. Now that you've accounted for the work up to this point, you should review all that you've studied. Here are some ways to do that.
   - ❑ Create a folderbook that covers the main ideas from each lesson. Instructions can be found in Appendix F.
   - ❑ Write a composition that reviews each lesson. See "Write About It!"—the composition worksheet and form in Appendix A.

     *(Note: If your science notebook is complete and you've done a good job with your Daily Reading Sheets, it should be fairly easy to get the facts you need for these summaries.)*

3. Show What You Know! — Complete the quiz on the next page to the best of your ability. After your teacher has reviewed your work, go back and use your book to answer anything you didn't know the first time through.

# Show What You Know!

Answer as many questions as you can without using your book or notes. You get **10,000** points for each correct response. After going through the quiz once with your book closed, have your teacher review your work. Then, open your book and try again. You get **5,000** points for each additional correct answer. So, **show what you know!**

1. How did the atmosphere come into being?

______________________________

2. What is Earth's atmosphere?

______________________________

3. Name three things that the atmosphere does for us.

______________________________

______________________________

4. Name two methods used to divide the atmosphere for study.

______________ ______________

5 Name the two layers used in the composition method.

______________ ______________

6. What layer of the homosphere is called the "weather layer"?

______________________________

7. Name the main layers according to the temperature method.

______________________________

______________________________________________

8. What instrument is used to measure atmospheric pressure?

______________________________________________

9. Who invented this instrument?

______________________________________________

10. What is the term for moisture in the atmosphere?

______________________________________________

11. Does sweat evaporate more quickly or more slowly when the air is very moist?

______________________________________________

12. List the main gases in the homosphere's composition and give their percentages.

______________________________________________

______________________________________________

First Attempt ______________
*(number of correct responses X 10,000)*

Second Attempt + ______________
*(number of correct responses X 5,000)*

**Total Number of Points** = ______________

# Unit Five
# Earth's Weather

## Unit Timeline

1805 The Beaufort Wind Scale is designed by Sir Francis Beaufort.

1857 Christophorus Henricus Didericus Buys-Ballot proposes the rule for the swirl direction of large storms and hurricanes.

1945 Antarctic explorers develop the wind chill index.

1959 The largest snowfall from one storm (189 inches) is recorded at Mt. Shasta Ski Bowl in California.

1974 Over a two-day period, 148 tornadoes strike 13 states.

2001 The wind chill index is modified.

## Unit Five Vocabulary

- cirrus
- clouds
- cold front
- condense
- convection
- cumulus
- cyclone
- Doppler radar
- dry snow
- eye
- freezing
- freezing rain
- Fujita scale
- Fujita, Theodore
- funnel cloud
- hail
- hurricane
- meteorology
- nimbus
- precipitation
- Saffir-Simpson scale
- seasons
- sleet
- snow
- stratus
- temperature
- thunder
- tornado
- tropical cyclones
- tropical depression
- tropical storm
- tropics
- typhoon
- warm front
- wall
- wet snow

## Materials Needed for This Unit

- Science notebook
- 3 photocopies of the "Checking It Out" form in Appendix A
- Photocopy of the unlabeled "Types of Clouds" illustration in Appendix A
- Photocopy of the "Weather Observation Chart" in Appendix A
- Photocopy of the world map in Appendix A (You can use the map you labeled in Unit One.)
- Materials for one or more scrapbook pages
- Drawing paper
- Jar — 1 quart or 1 pint
- Funnel
- Sandwich- or quart-size plastic bag
- Measuring cup
- Ice cubes
- 2-liter bottle, empty and clean, with label and cap removed
- Sharp scissors
- Duct tape
- Permanent marker and colored pencils
- 2 2-liter bottles, empty and clean, with labels and caps removed
- Modeling clay
- Small bag metallic confetti (This can be purchased at a craft store or card shop.)
- Globe or atlas
- Materials for Unit Wrap-Up (See page 201.)

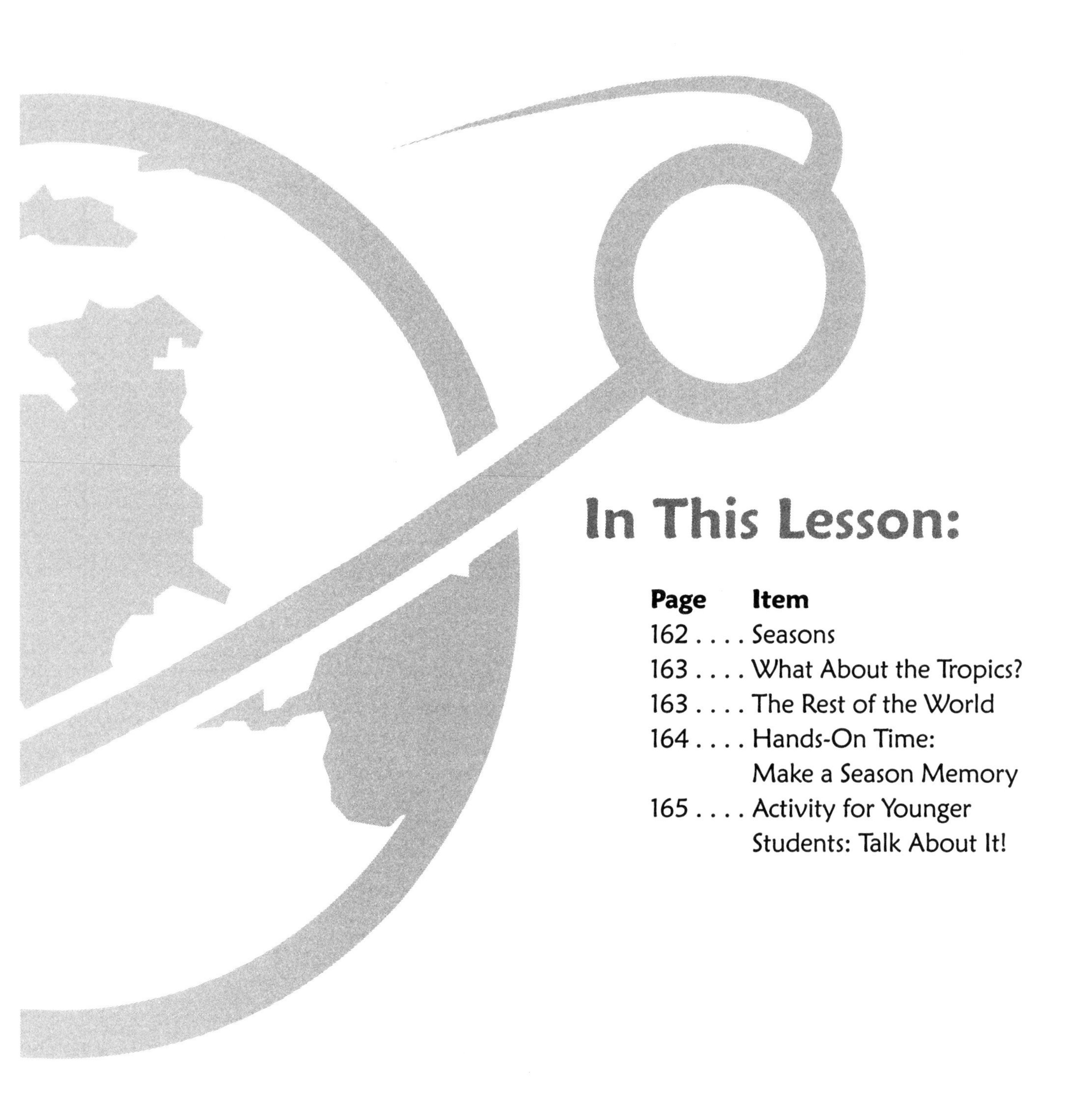

## In This Lesson:

Lesson 16

# INTRODUCTION TO WEATHER AND SEASONS

## Teaching Time:

## Be Weather-wise

Throughout your study of our Earth, you have had the opportunity to learn many things about Earth and what goes on around you. You've also had the opportunity to see how all of this fits in with the Bible and with what God tells us about Himself. God is a magnificent Creator and Designer. Truly, we cannot fully understand our world without considering God and His plan for us all. Recently, I heard a major in the United States Air Force say, "True science is discovering how God designed our world." This man is a pilot and, like me, the more he learns, the more his faith in God and the Bible is strengthened.

"For He [the LORD] commands and raises the stormy wind, which lifts up the waves of the sea." ( Psalm 107:25)

### Weather All Around Us

Every day in every part of the world at all times there is weather. Just think about that for a moment. Weather is always happening . . . everywhere! Since that is the case, we can benefit

**temperature**
How hot or cold the air is.

**seasons**
Changes in weather patterns based on Earth's tilt toward the sun.

**precipitation**
Water that falls to Earth in the form of rain, snow, sleet, hail, or freezing rain.

from learning about weather, how it happens and why. That is the main thing we're going to talk about in this unit.

Weather is made up of many different elements. First, there's **temperature**—how hot or cold the air is. The temperature of the air around us is a major factor in our weather. For instance, if it's warm, it won't snow. Temperature, as you will learn in lessons to come, affects how clouds form. Temperature, as I'm sure you know, is a factor in our seasons. **Seasons** are the changes in our weather patterns based on Earth's tilt toward the Sun.

Another factor in weather is **precipitation**, which is water that falls to Earth in the form of rain, snow, sleet, hail, or freezing rain. We will study more about precipitation in another lesson.

Weather can be calm, or it can be violent. Violent weather can be in the form of thunderstorms, tornadoes, and hurricanes. Some parts of the world experience typhoons and monsoons. One major factor in some of this weather is air pressure, which you have already studied if you have done the unit on the atmosphere.

## Seasons

The United States, like many other countries, has four seasons. Can you name them on your own? There's winter, spring, summer, and autumn (sometimes called fall). These four seasons each have different traits. Some people prefer one season over the other, but that is determined by their own likes and dislikes. I like something about each of the seasons. What about you?

As I mentioned earlier in the lesson, seasons are caused by Earth's tilt toward the Sun. Do you remember what you learned at the beginning of this book about the axis of Earth and how Earth spins on its axis? Think back also to what you learned about Earth's revolving around the Sun. Earth is tilted on its imaginary axis, constantly spinning, or rotating, causing our days and nights and at the same time, it is in the process of revolving around the Sun. It takes 365¼ days for Earth to make one complete revolution around the Sun. For half of this time, or for six months, the Northern Hemisphere is closer to the Sun than the Southern Hemisphere. This means the Northern Hemisphere is having its warmer weather. For the remainder of the year, the Southern Hemisphere is closer to the Sun, having its warmer weather. Therefore, you could expect that if you traveled in January from the United States to Brazil, the weather would not be the same in both places. In other words, we're having winter while they're having summer. The angle of Earth toward the Sun is also responsible for our longer days in summer and shorter days in winter.

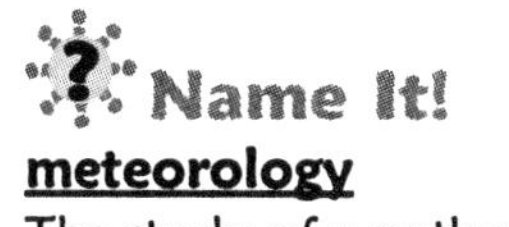

**meteorology**
The study of weather.

**tropics**
The regions around the equator, characterized by consistently warm temperatures and warm ocean waters.

## What About the Tropics?

The area of Earth that is near the equator is referred to as the **tropics**. The distance from the tropics to the Sun changes very little, which means the temperature and weather conditions in the tropics are fairly constant. Once I was able to go to Hawaii in October. It was about 80 degrees every day. When I landed in St. Louis, Missouri, after my trip, it was snowing!

## The Rest of the World

Not all parts of the world experience four seasons. In fact, even in the United States, some states like Florida have very little winter but a lot of summer. Seasons in other parts of the world are described as wet or dry. In some places the rain falls mostly during one part of the year, which is called the wet season.

Weather and the seasons are major factors in our lives. We have to deal with their impact on our lives at all times. Learning more about them may help you make decisions as you go through your life. I hope you enjoy all the things we study over the next few weeks. The study of weather is called **meteorology**. If you are particularly fascinated by weather, you might consider becoming a meteorologist when you grow up.

Additional Notes

## Hands-On Time:
# Make a Season Memory

## Featured Activity

### Objective
To consider the different seasons where you live and how they affect you, and to report on them.

### Instructions
Today's Hands-On Time is planned as a writing assignment. However, I am including some other options. With your teacher, choose which one you should do.

For those of you who are very creative and like to write stories, you can write a story about a boy or girl experiencing your favorite season. Be sure to include information about the temperature and what activities he or she enjoys during that time of year. A different take on the same idea is to write about his or her *least* favorite season.

For those of you who are more technical and do not like to make up stories, you can write a report on the seasons where you live. Please be sure to include facts about the temperature and rainfall/snowfall. If your area is likely to have violent weather, you might include that information as well.

If you prefer to do visual art, create a scrapbook page that shows you enjoying your favorite season. Perhaps it's a snow day or a day at the beach. Be sure to create a title for your page and add some journaling to tell what is happening in the pictures and what the weather is like.

Another option is to draw a picture of yourself, a friend, or your family outside in your favorite season. Think about the trees. Are there leaves on the trees in winter? Would a tree have a green top then? What would the grass look like? Make sure your picture looks like the season you are drawing.

**Additional Notes**

## Activity for Younger Students: Talk About It!

Talk about the different seasons with your teacher. Describe what you like about each season and why. Do you have a favorite season? Which is it? What do you like to do the most during that time of the year? What do you see in nature during that time?

*Note to Teacher*: Remember, conversation is a major learning tool and relationship builder. You can help your student learn to love his or her schoolwork by having interesting conversations about the things your student is studying. It is also a very good way to review what your student has read and make sure the concepts are understood.

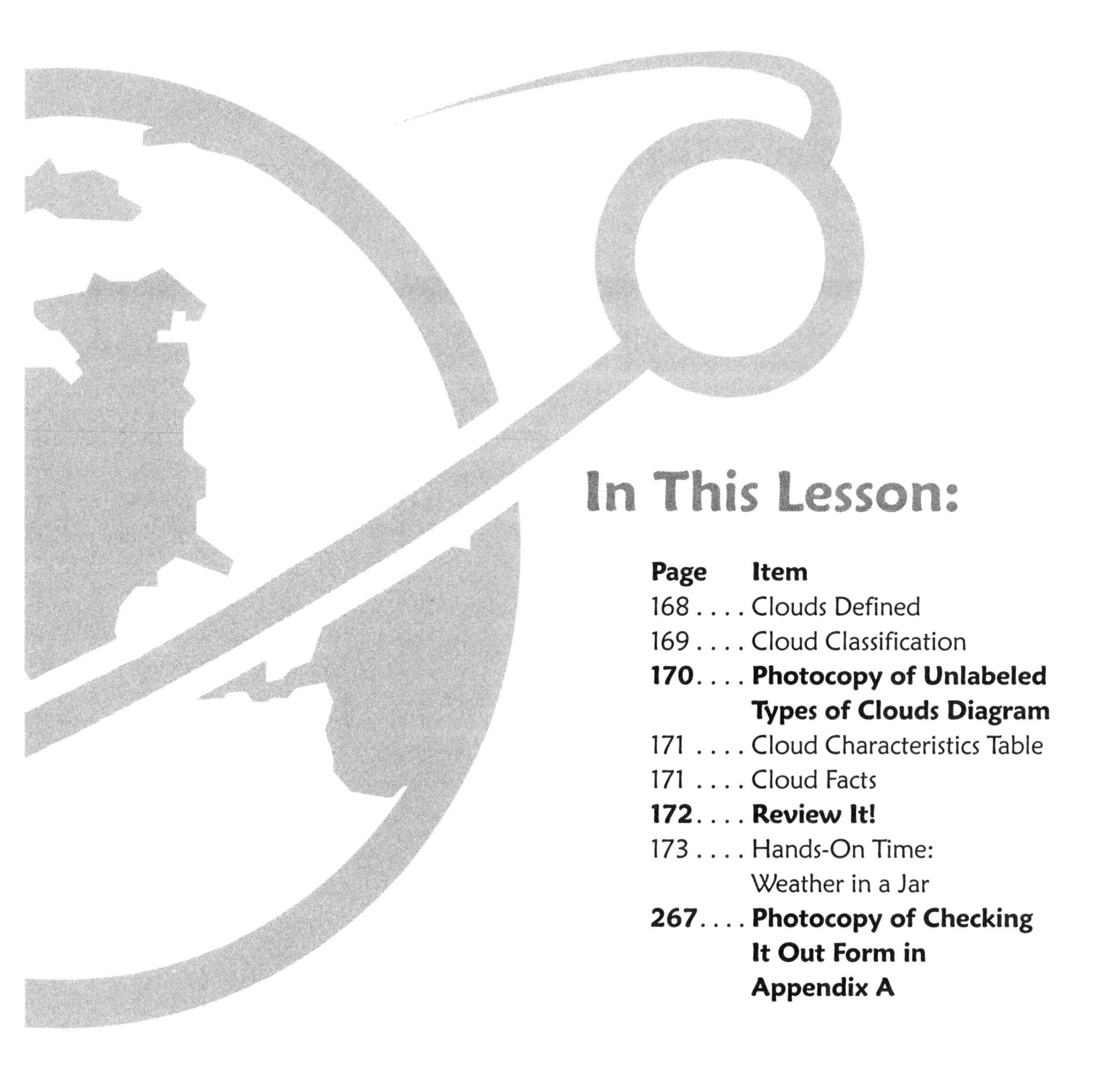

# In This Lesson:

Lesson 17

# CLOUD FORMATION

## Teaching Time: All About Clouds

Going on vacation is one of my favorite things to do. I don't mean the amusement-park type of vacation, though that is fun. Instead, I mean the sit-by-the-lake type of vacation. In fact, that's what I'm doing right now. I'm sitting by a quiet lake, looking at all the beauty and listening to the many wonderful and often, strange sounds. This type of vacation tends to turn toward work for me. Why? Well, it's here that I have time to be still and think about the work of God in creation. It's then that I feel I must write and share with you the things I believe God is teaching me.

Scripture

"By His [God's] knowledge the depths were broken up, and clouds drop down the dew." (Proverbs 3:20)

You may wonder what all of this has to do with clouds. I'm glad you asked! When I'm on vacation at the lake, I love to watch the sky. This helps my husband and me figure out what the day holds (or at least, the next several hours). A major part of this is watching the clouds and knowing what each type of cloud might mean when it comes to weather. Do you know that there are different types of clouds? Do you know that by studying

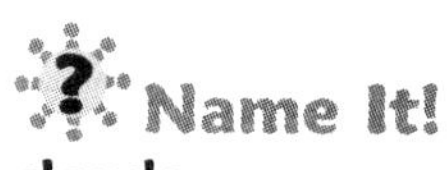

**Name It!**

**clouds**
Collections of very small water droplets and ice crystals that form when water is evaporated from the ground and then returns to liquid form in the air.

**condense**
To change back to liquid form.

clouds you can learn a lot about the upcoming weather? Like almost everything, there's more to clouds than meets the eye. Let's get busy learning!

## Clouds Defined

You probably know already that clouds are made up of water and that they are responsible for rain. That is a good start in understanding clouds, but it is just the beginning. **Clouds** are collections of very small water droplets and ice crystals that form when water is evaporated from the ground and then **condenses**, returns to liquid form, in the air.

You know from our lesson on humidity in another unit that there is always water present in the air in the form of water vapor, which is invisible.

- As air moves about in the troposphere, the lower part of the atmosphere, some air rises, cooling as it goes.

- As the air cools, it can reach the point where the water condenses.

- Eventually, the air is holding as much water as it can for the temperature, or is saturated. In this case, we get precipitation, which is your next lesson.

Here's an everyday example. Have you ever had a glass of cold lemonade, or any other beverage, and seen the glass get foggy? That fog is a cloud, of sorts, forming on the outside of your glass due to the cold temperature of the glass. Eventually, you may see water droplets on the glass. Those droplets are condensation, or moisture, from the air that has cooled to a point where it is liquid. As with clouds, the droplets on the glass can combine to make larger droplets, which can end up "raining" down the glass. I've heard people say the glass is "sweating." Our Hands-On Time will teach you more about this concept.

# Cloud Classification

Clouds can be identified and grouped according to their shape, size, and height in the atmosphere. To help you understand the differences between clouds, we're going to take a look at some Latin. You see, the names of the different types of clouds are Latin, and the translation will help you understand the cloud. Let's look at the table below.

| Cloud Name (Latin) | English Translation |
|---|---|
| Cumulus | heap |
| Stratus | layer |
| Cirrus | curl of hair |
| Nimbus | rain |

Using this table, can you identify each type of cloud in the illustration provided on the next page? (Check your answers against the labeled version of this illustration in Appendix A.)

In addition to the cloud types listed above, cloud names can also have words connected that describe the height in the atmosphere where the clouds are formed. For instance, altocumulus clouds are mid-level, puffy clouds. The next table gives you the name of the cloud type, its height in the atmosphere, and what its main features are. Read through this table to learn more.

## Name It!

**cumulus**
Clouds formed by rapidly rising air, extending up to 39,000 feet and higher; name translated means "heap."

**stratus**
Clouds formed by slowly rising air; can occur as high as 20,000 feet and last for days. Name translated means "layer."

**cirrus**
Thin, white clouds containing ice crystals; found at 20,000 feet and higher; name translated means "curl of hair."

**nimbus**
Lower, dark clouds that contain water; name translated means "rain."

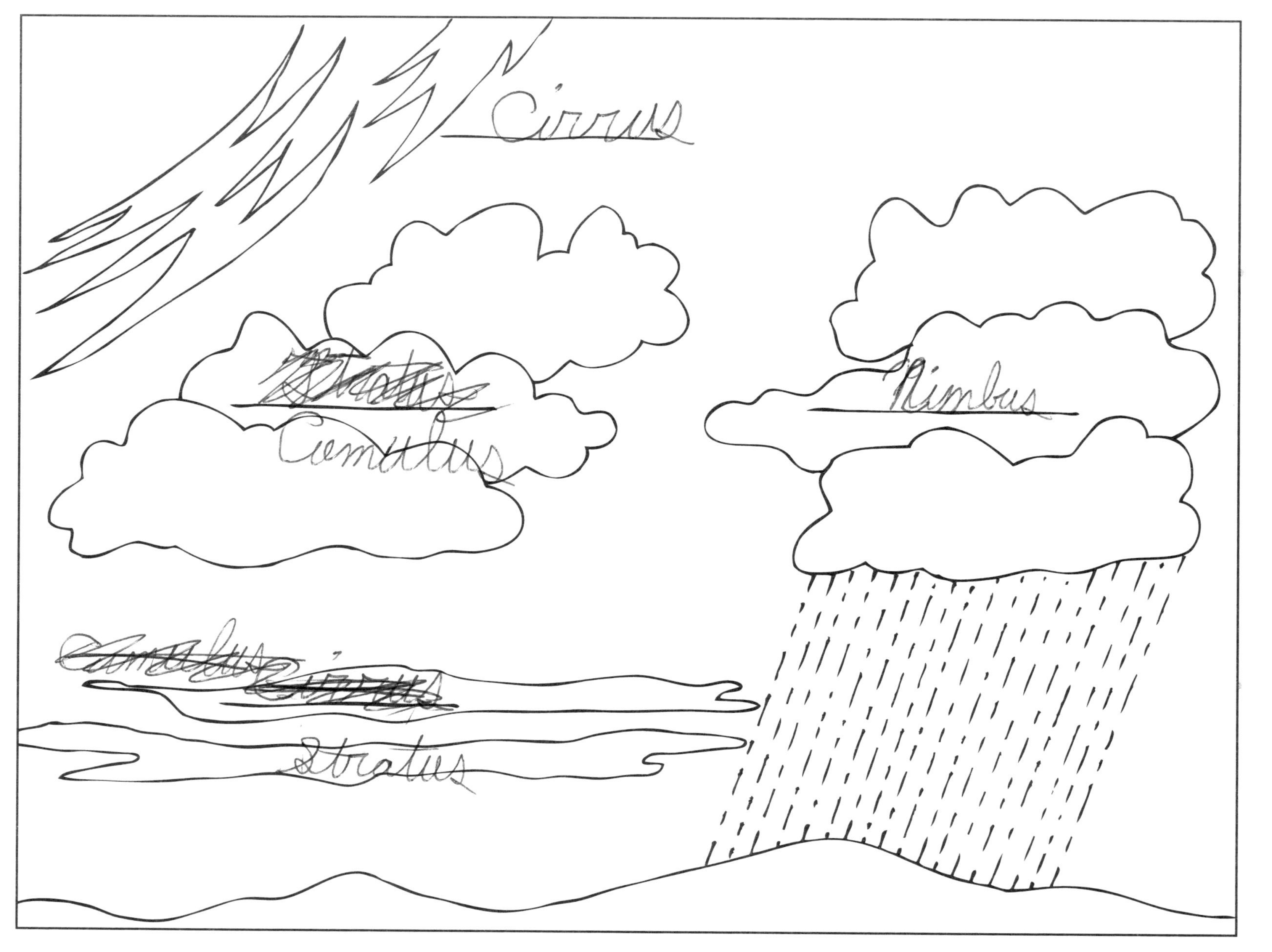


Types of Clouds

Additional Notes

## Cloud Characteristics

| Height *(in feet)* | Cloud Names | Features | Contain |
|---|---|---|---|
| >39,000 | Cumulonimbus, fair-weather cumulus | Anvil shaped, vertically developed | Thunderstorms, release much energy |
| 20,000 & up | Cirrus, cirrostratus | Thin, white | Ice crystals |
| 6,500 - 20,000 | Altocumulus, alto-stratus | White, often puffy | Water droplets / ice crystals |
| <6,500 | Nimbostratus, stratocumulus | Overcast, thick, puffy | Water droplets unless very cold, then ice and snow |

As I said, when I'm on vacation at the lake, I watch the sky. I pay attention to the changes in the clouds. When I see thin, scattered wisps of clouds, I don't worry. It's out in the boat for my family! However, if I look out and see clouds building upward and flattening out at the top, like an anvil, I take caution and suggest a board game instead.

# Cloud Facts

- They're formed by moisture in rising air, cooling and condensing.
- Cumulus clouds are formed by rapidly rising air, last a few hours and cover a small area.
- Stratus clouds are formed by slowly rising air, can last for days and cover a wide area.
- Darker clouds are dark because they contain more water and therefore, block more sunlight.

We will add to this information in our next lesson, so hang on until then.

**Additional Notes**

# Review It!

*May be done orally or written.*

1. Clouds are collections of very small water droplets and ice crystals that form when water is evaporated and then returns to liquid in the air.

2. Based on the meanings of the cloud names, which clouds are

   a. Puffy? ______________________

   b. Rainy? ______________________

   c. Wispy? ______________________

3. ______________ clouds are very high clouds that are anvil shaped and contain thunderstorms.

4. Since stratus means "layer" and cumulus means "heap," what do you think stratocumulus clouds look like?

   ______________________________

5. How do you know that nimbostratus clouds bring rain?

   ______________________________

   ______________________________

6. What do dark clouds have that make them so dark?

   more water

## Hands-On Time:
# Weather in a Jar

## Featured Activity

### Objective
To create a cloud, and precipitation that we can observe.

### Materials
- Jar – 1 quart or 1 pint
- Funnel
- Sandwich- or quart-size plastic bag
- Measuring cup
- Ice cubes
- Hot and cold water
- Photocopy of the "Checking It Out" form in Appendix A

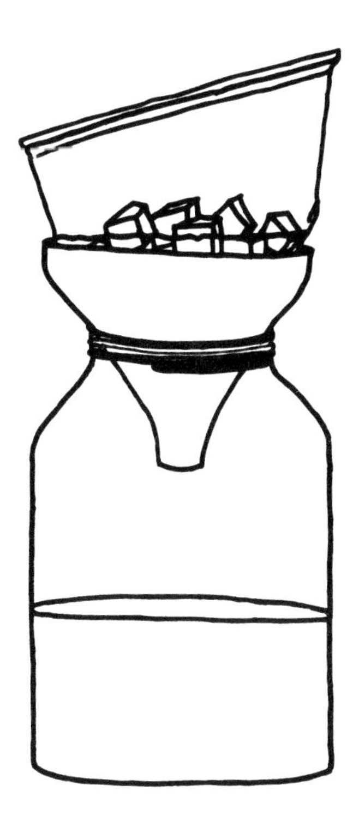

### Instructions
(*See the illustration accompanying this activity.*)

1. Measure out one cup of hot tap water. Pour into jar.
2. Record what you see.
3. Set funnel on the jar opening.
4. Line the funnel with the plastic bag and fill it with ice. Add ¼ cup cold water.
5. Carefully observe all the changes in your jar.

Additional Notes

6. Record all your observations on your "Checking It Out" form.
    - List immediate observations.
    - Record observations every 5 minutes.
    - The effect of the ice: Did you see any changes? Did you see condensation? Did you see precipitation? What caused the changes?

7. Complete your "Checking It Out" form.

8. Add to your science notebook.

The ice created a cooler temperature in the upper atmosphere of your jar. As it cooled the vapor from the warm water, you may have seen fog, or a cloud. (*Note*: It probably did not look like the clouds you see up in the sky.) Next, you should have seen water droplets form on the jar. This is the condensation. As the water droplets pooled together, you probably saw precipitation. Now you have created your own weather in a jar!

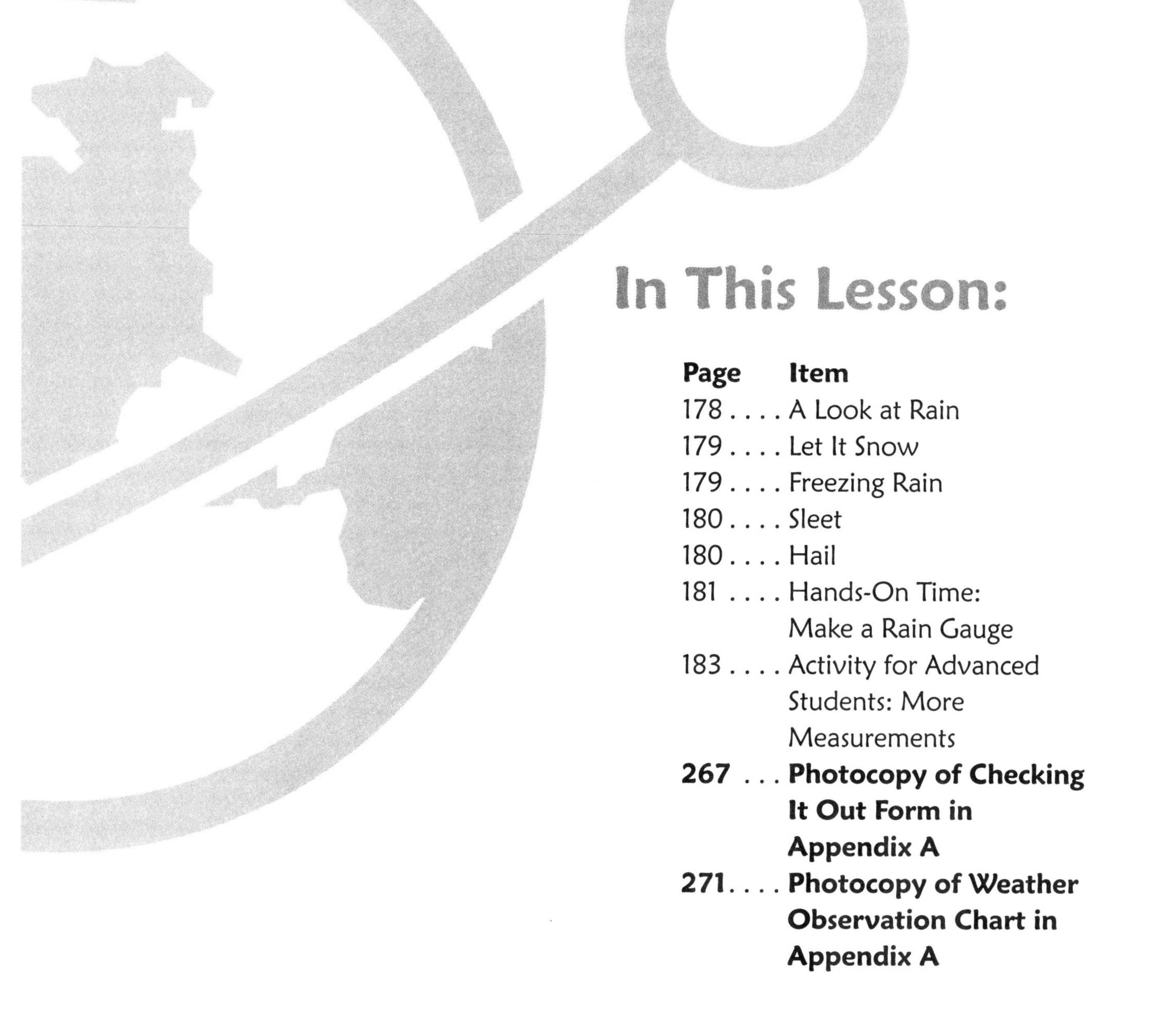

# In This Lesson:

**Page** **Item**

Lesson 18

# PRECIPITATION

Teaching Time:

## Rain, Snow, Sleet, and Hail

Now that we have learned a few things about clouds, it seems only natural to study the water that falls to Earth from those clouds. This is called **precipitation**, and it comes in several forms: rain, snow, sleet, hail, and freezing rain. As we begin today's lesson, I'd like you to spend a few moments thinking about what types of precipitation you have seen in your life. You see, not all places on Earth experience all types of precipitation. I have seen all types, and I can tell you that I pretty much like any type of precipitation as long as there's not too much of it for too long.

**Scripture**

"Praise the LORD from the earth, you great sea creatures and all the depths; fire and hail, snow and clouds; stormy wind, fulfilling His word . . . " (Psalm 148:7–8)

**Name It!**

**precipitation**

Water that falls to Earth in the form of rain, snow, sleet, hail, or freezing rain.

### Types of Precipitation

As you've already learned, clouds are formed by water vapor condensing as warm air rises and cools. The type of precipitation

**cold front**
Cool air pushing in on warm air.

**warm front**
Warm air slowly rising as the cooler air sinks, causing condensation to take place.

that can result depends on how fast the air rises and what is going on in the rest of the troposphere in that area. (Do you remember what the troposphere is? It's the lower layer of the atmosphere where weather takes place.) As I mentioned in the opening paragraph, precipitation can be in the form of rain, snow, sleet, hail, and freezing rain. Which type do you prefer? Which type have you found to be the most inconvenient?

## A Look at Rain

I'm sure you've experienced rain many times in your life. You may have noticed that sometimes rain falls steadily for days and other times it comes in huge downpours that last only a few minutes. In the southern United States, we call those quick downpours "gully washers." Have you ever wondered what causes the difference? Gully washers, as I like to call them, are caused by rapidly rising warm air. They are common in the summer due to the high temperatures and high humidity but can happen because of a **cold front**, cool air pushing in on warm air. In summer, the air on the ground heats up and rises very quickly. If there is a lot of water vapor present, meaning high humidity, big cumulus clouds form. All the tiny water droplets quickly gather together, forming huge raindrops that are released in one great *whoosh* as downward currents in the clouds force them down.

Long-lasting rains are a different matter. They come from nimbostratus clouds that form as a **warm front** moves into an area. The warm air slowly rises as the cooler air sinks. As the warm air rises and cools, condensation again takes place, often resulting in rain. The reason for the slow movement is that cool air moves more slowly than warm air. The entire process is gentler than the weather systems that cause heavy rain, and the precipitation is also gentler. Because everything is moving slowly and nimbostratus clouds cover large areas, the rain can last for a long time, often days.

## Let It Snow

You just learned that long, steady rains come from nimbostratus clouds resulting from warm fronts. What I didn't tell you is that the precipitation from these clouds is not always in the form of rain. In fact, if you think back to our lesson on clouds, you may recall that I told you snow can also fall from nimbostratus clouds. If the air temperature is around the freezing point, you just might get a snowfall instead of rain.

**Snow** is many ice crystals grouped together that fall to the ground in the form of snowflakes. The ice crystals are not tightly packed but rather cushioned by air, forming the snowflakes. Usually, it takes 10 inches of snow to equal 1 inch of rain. In other words, if you melted 10 inches of snow, you would end up with 1 inch of water. However, if the air is very cold, the snowflakes are lighter, giving a **dry snow**. This type of snow is hard to pack together and is very powdery. **Wet snow** results when the air is just below freezing. Wet snow packs together well and makes good snowballs and snowmen. The reason for the difference is that air that is well below freezing holds less moisture than air that is near freezing. **Freezing** is 32 degrees Fahrenheit.

## Freezing Rain

I mentioned freezing rain and I wonder if you know what freezing rain is. **Freezing rain** is rain that falls from the clouds and freezes when it comes in contact with the ground or things on the ground. Freezing rain happens when the ground is below 32 degrees Fahrenheit, but the air temperature is higher. Freezing rain forms layers of ice on all it touches. It weighs down power lines and tree branches, often causing them to collapse under the weight. It forms sheets of ice on the roads, making them very dangerous to drive on. Freezing rain is pretty to look at because it makes everything look like glass, but it causes a lot of damage.

**snow**
A form of precipitation made up of many ice crystals grouped together that fall to the ground as snowflakes.

**dry snow**
Snow that occurs in very cold air when the snowflakes are lighter.

**wet snow**
Snow that occurs when the air is just below freezing; good "packing" snow.

**freezing**
32 degrees Fahrenheit.

**freezing rain**
Rain that falls from the clouds and freezes when it comes in contact with the ground or things on the ground. Occurs when the ground is below 32 degrees Fahrenheit, but the air temperature is higher.

**sleet**
Begins as rain but becomes ice as it is falling to the ground. Occurs when the air temperature is above freezing but decreases closer to the ground.

**hail**
Water droplets that have been tossed up and down in violent cumulonimbus clouds freezing to ice, collecting more water, freezing more, over and over until balls of ice are formed.

## Sleet

Many people confuse sleet and freezing rain. **Sleet** begins as rain but becomes ice as it is falling to the ground. Sleet occurs when the air temperature is above freezing but decreases closer to the ground. Remember that freezing rain doesn't freeze until it comes in contact with the ground or things on the ground. Sleet, on the other hand, becomes frozen as it passes through the air on the way to the ground. Freezing rain forms a solid sheet of ice, but sleet is a bunch of pieces of ice that accumulate more like snow.

## Hail

One last form of precipitation is hail. **Hail** is water droplets that have been tossed up and down in violent cumulonimbus clouds freezing to ice, collecting more water, freezing more, over and over until balls of ice are formed. I have personally seen hail as small as peas and also as large as golf balls. The air inside cumulonimbus clouds is full of energy. There are rapid updrafts and downdrafts of air inside these clouds. As these clouds build, they climb higher and higher. When hail forms inside these huge clouds, it is tossed up and down along with the rising and sinking currents of air. The more rapid the air movement, the larger the hail. Hail can be very dangerous, as can the other weather that comes along with it. You'll learn more about that in another lesson.

Precipitation is a key element in the hydrologic cycle. It brings the water that has been evaporated back to Earth, watering the ground and plants and replenishing our lakes and streams. God has designed our Earth to have all these varied types of precipitation. The most famous rainfall ever is recorded in the Book of Genesis and is known as Noah's Flood. God promised then that He would never destroy all the life on Earth again with a flood. He takes care of us and our Earth with the precipitation He sends.

## Hands-On Time:
# Make a Rain Gauge

## Featured Activity

### Objective
To observe how much rain you are receiving and practice recording information.

### Materials
- 2-liter bottle, empty and clean, with label and cap removed
- Sharp scissors
- Duct tape
- Permanent marker
- Photocopy of the "Checking It Out" form in Appendix A
- Photocopy of the "Weather Observation Chart" in Appendix A

### Instructions
This is a simple way to measure the amount of rain that you get at any one time. Using the instructions below, build a rain gauge and monitor the results day by day for at least one week. If no rain falls during that time, please continue monitoring. (*See the accompanying illustration for guidance.*)

1. Cut around the top third of the bottle to remove it.

2. Turn the top portion upside down into the bottom portion. It should resemble a funnel.

3. Tape around the edges to hold the "funnel" in place.

Additional Notes

4. Add water to the container until any ridges in the bottom are filled and there is an even starting point for measuring. Put a mark here on the outside of the container. You will need to have water at this level before collecting any rainwater.

5. From this starting point, measure off and mark 1-inch increments and label them. Make a smaller mark at the halfway point between inches. You may even want to mark the ¼-inch points as well, for more accurate readings.

6. Place your rain gauge away from houses, trees, and anything that might block the rain. Be sure to place it securely so it will not be blown over!

7. After a rainfall, carefully move your gauge to a very flat surface and record the measurement in a form like the one shown here. REMEMBER: If you did not have water in your container up to the starting point, your reading will not be entirely accurate. Therefore, check your gauge before rain is expected and replace any evaporated water.

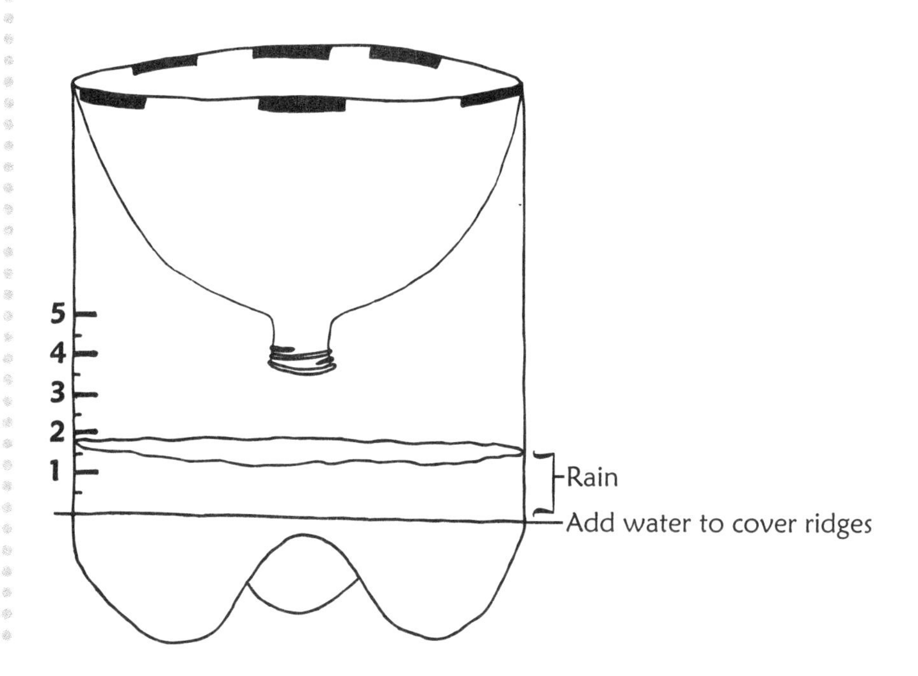

### Rainfall Measurements

| Date: | Date: | Date: | Date: | Date: |
|---|---|---|---|---|
| Rainfall Amt. inches | Rainfall Amt. inches | Rainfall Amt. inches | Rainfall Amt. inches | Rainfall Amt. inches |

8. Complete a "Checking It Out" form and place it in your science notebook.

9. NOTE: After a rainfall, you will need to empty all the water except that which goes to the starting point.

## Activity for Advanced Students: More Measurements

1. In addition to building a rain gauge, use the barometer, psychrometer, and thermometer from your other experiments to measure the weather each day.

2. Use the "Weather Observation Chart" in Appendix A to record your findings.

3. Observe how your readings compare with the actual weather each day.

4. Add this chart to your science notebook.

### Additional Notes

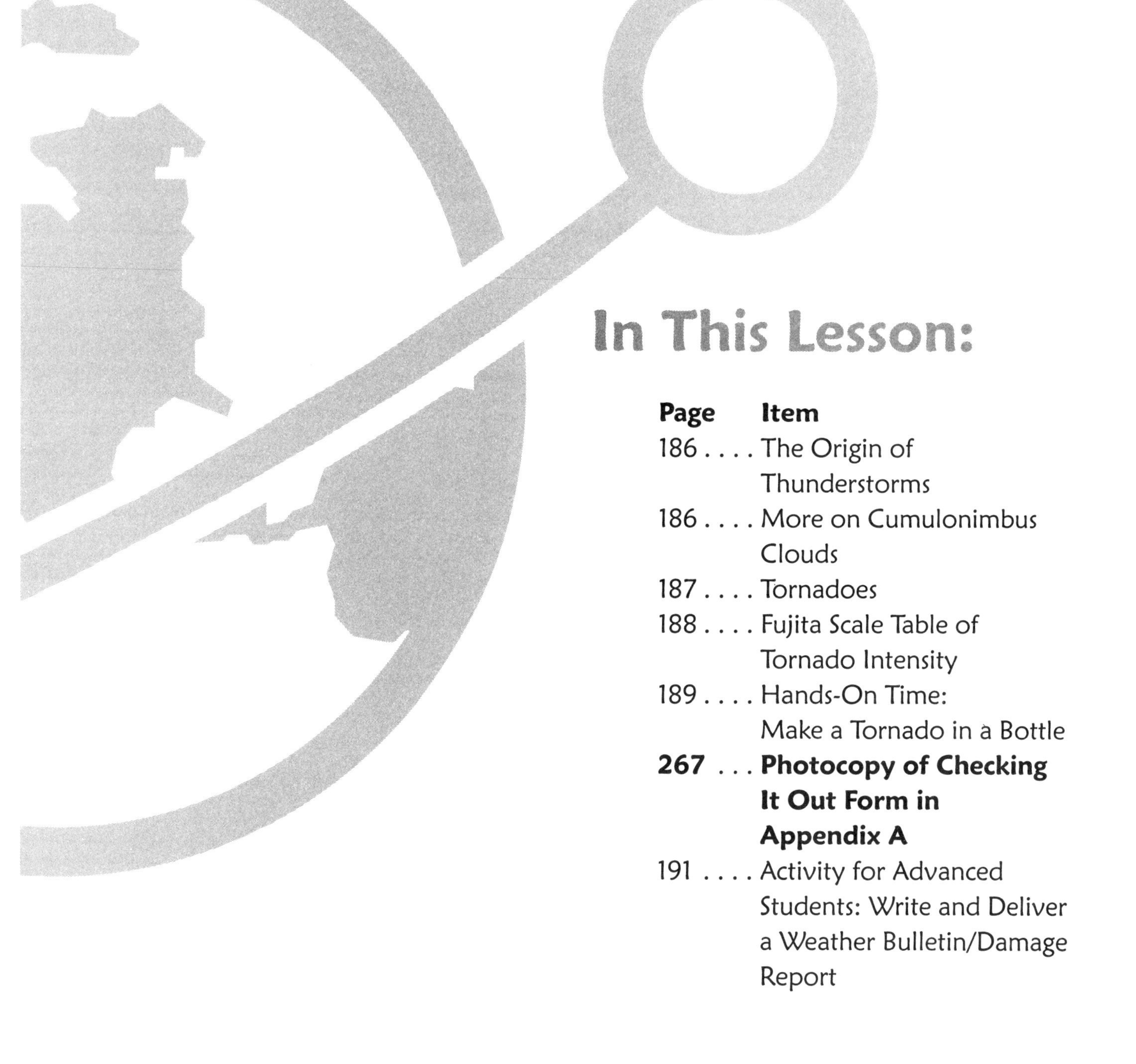

# In This Lesson:

Lesson 19

# THUNDERSTORMS AND TORNADOES

## Teaching Time: When Weather Turns Violent

Today we're going to talk about two types of weather that can be quite violent: thunderstorms and tornadoes. It's only fair to you that I make a confession right here and now. From the time I was a young girl until I was in my thirties, I was absolutely terrified of both thunderstorms and tornadoes. I don't know how the fear began, but I know that the more I thought about it and dwelt on it, the worse the fear became. As I write this book, I am into my forties by a few years and no longer am terrified. You may wonder what has made the difference. Well, first of all, weather forecasting is very good these days. There is usually advance notice of impending danger from storms, giving those of us affected time to prepare. Second, and most important, my faith in God is stronger than it used to be. If I could impart any one thing to you, it would be to learn to trust God more. God is in control and nothing happens to you apart from His knowl-

 **Scripture**

"He causes the vapors to ascend from the ends of the earth; He makes lightning for the rain; He brings the wind out of His treasuries." (Psalm 135:7)

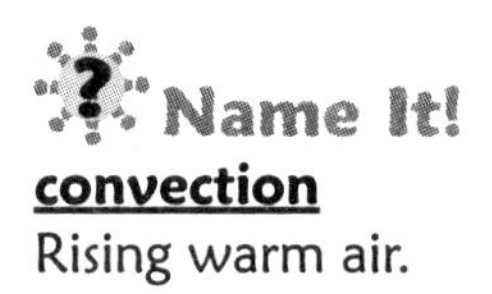


**convection**
Rising warm air.

edge. Furthermore, He is able to handle anything that may come your way. Now . . . back to our lesson.

## The Origin of Thunderstorms

Our lessons in weather, up to this point, have covered clouds and precipitation. Believe it or not, having done these lessons, you are well prepared for our discussion on thunderstorms. Thunderstorms result from cumulonimbus clouds. Do you remember what the root words *cumulus* and *nimbus* mean? I hope that you said "heap" and "rain." Cumulonimbus clouds are clouds that form in heaps and contain rain. More specifically, they are formed from rapidly rising warm air that is surrounded by cooler air above. The rising warm air is referred to as **convection** (kun **vek** shun). Sometimes the warm air is forced up by mountain ranges, sometimes by a cold front moving in, and sometimes by a low-pressure system moving in. The greater the difference in temperatures of warm air meeting cool air, the more violent the resulting weather can be.

## More on Cumulonimbus Clouds

Cumulonimbus clouds contain a great deal of energy. The air inside moves up and down, moving the moisture inside up and down with it. As you know, this process sometimes creates hail. Cumulonimbus clouds begin as cumulus clouds. As the warm air rises and encounters more and more cool air, cumulus clouds form. These cumulus clouds that are building upward, or vertically, combine and grow into a massive storm. The tops of these clouds tend to flatten out on the top from wind shear across the upper level of the atmosphere. As the moisture inside builds, creating rain, they become cumulonimbus.

So you can see that the ingredients for cumulonimbus clouds are:

- Rapidly rising warm air
- Large amounts of water vapor

The energy within these large clouds can create lightning, thunder, hail, and tornadoes, in addition to large amounts of rain. **Thunder** is the sound created by the expansion of the air that is suddenly heated by lightning. Lightning can heat the air up to 54,000 degrees Fahrenheit! Thunder happens at the same time as lightning, but we see lightning before we hear thunder because light travels faster than sound.

## Tornadoes

Tornadoes can happen when the winds inside the storm cloud begin to blow around a center of very low air pressure. The spinning of the air and the water vapor inside creates a **funnel cloud**. The funnel cloud comes down out of the storm cloud, and if it actually makes contact with the ground, it becomes known as a **tornado** (tor **nay** doe). Tornadoes can move at speeds up to 70 miles per hour. They can create wind speeds from 75 to 300 miles per hour.

Tornadoes are quite common in the United States. They also occur in many other places in the world. Tornadoes are often rated based on the amount of damage they cause. Wind speeds are also a factor in their rating. The scale used to rate tornadoes is called the **Fujita scale** and was developed by **Theodore Fujita** in 1971. Tornadoes are ranked from F-0 (little to no damage) up to F-5 (very destructive). See the chart provided in this lesson to learn more about the Fujita scale.

As I mentioned at the beginning of our lesson, weather prediction has come a long way. One of the most helpful tools in predicting oncoming severe weather such as thunderstorms and tornadoes is **Doppler radar**. Doppler radar allows meteorologists to see through computer images what is going on inside storms. These images are often broadcast on television— allowing viewers time to prepare for storms, mainly by getting themselves and their families out of harm's way. During the time I lived in Oklahoma, I was able to overcome my great fears. Interestingly, this is where I experienced the most severe

### Discovery Zone

Learn more about tornadoes at www.factmonster.com under "Science."

### Name It!

**thunder**
Sound created by the expansion of the air that is suddenly heated by lightning.

**funnel cloud**
Spinning of the air and the water vapor inside.

**tornado**
Funnel cloud that comes down out of a storm cloud and actually makes contact with the ground.

**Fujita scale**
Scale used to rate the strength of a tornado.

**Theodore Fujita**
Developed the Fujita scale in 1971.

**Doppler radar**
Weather-predicting tool that allows the inside of a storm to be seen through computer images.

weather of my life. God used something as simple as my watching my husband teach our children about the storms and how God is in control to calm my unreasonable fears and teach me to trust Him more. I am so grateful!

### Fujita Scale of Tornado Intensity

| F-Scale Number | Intensity Phrase | Wind Speed | Type of Damage Produced |
|---|---|---|---|
| F-0 | Gale tornado | 40–72 mph | Tree/sign damage. |
| F-1 | Moderate tornado | 73–112 mph | Hurricane-force winds; roof damage; can move cars |
| F-2 | Significant tornado | 113–157 mph | Considerable damage; roofs can separate; large trees can be uprooted. |
| F-3 | Severe tornado | 158–206 mph | Well-built homes significantly damaged; forests destroyed. |
| F-4 | Devastating tornado | 207–260 mph | Well-built homes destroyed; cars thrown. |
| F-5 | Incredible tornado | 261–318 mph | Strong homes moved far from foundations and disintegrated; steel-reinforced concrete structures badly damaged. |
| F-6 | Inconceivable tornado | 319–379 mph | Very unlikely. Difficult to distinguish from damage that would accompany the F-4/F-5 winds that would also occur. |

## Hands-On Time:
# Make a Tornado In a Bottle

## Featured Activity

### Objective
To simulate the spiral movement of a tornado and the flying debris.

### Materials

- 2 2-liter bottles, empty and clean, with labels and caps removed
- Modeling clay
- Duct tape
- Water
- Small bag metallic confetti (This can be purchased at a craft store or card shop.)
- Photocopy of the "Checking It Out" form in Appendix A

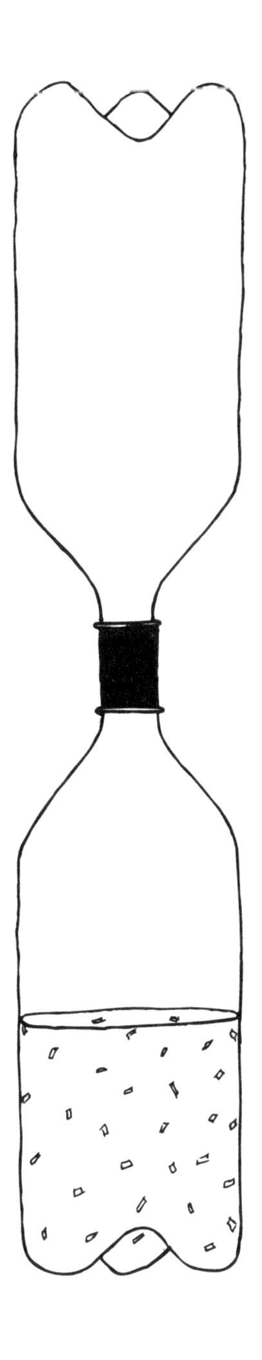

Additional Notes

### Instructions

(*See the illustration accompanying this activity.*)

1. Make a roll about the thickness of your little finger with the modeling clay.
2. Press it around the opening of one of the bottles.
3. Fill the bottle with water about one-half to three-quarters full.
4. Add confetti to simulate "debris."
5. Invert the second bottle on top of the first bottle.
6. Press it down on the modeling clay to help seal the two bottles.
7. Holding the bottles straight, wrap enough duct tape around the necks to seal firmly.
8. Over a sink or outdoors (in case the seal fails), turn the two bottles upside down.
9. Watch the pattern of the water and the "debris."
10. Consider: If this was much larger than your house and the debris consisted of cars and roofs, where would you be safe? Would it be wise to be outside or near a window watching?
11. Discuss with your teacher the best plan of action in case of a tornado. Where in your home should you go?
12. Complete a "Checking It Out" form and add it to your science notebook.

# Activity for Advanced Students: Write and Deliver a Weather Bulletin/Damage Report

Additional Notes

## Objective

To consider the impact of a tornado in a community.

## Instructions

1. Choose a town near you, or the one you live in, as the site of a tornado strike.

2. Decide where the tornado hit and what was damaged as a result.

3. Rate the tornado, using the Fujita scale, according to the damage produced.

4. Imagine that you must present the information on television or radio. Write a news bulletin that accurately describes the event (time, location, severity, and resulting damage). Be sure to include the tornado's rating according to the Fujita scale.

5. Once your bulletin is written, deliver this presentation to your family and friends. You might even want to make a videotape.

6. Place your bulletin in your science notebook.

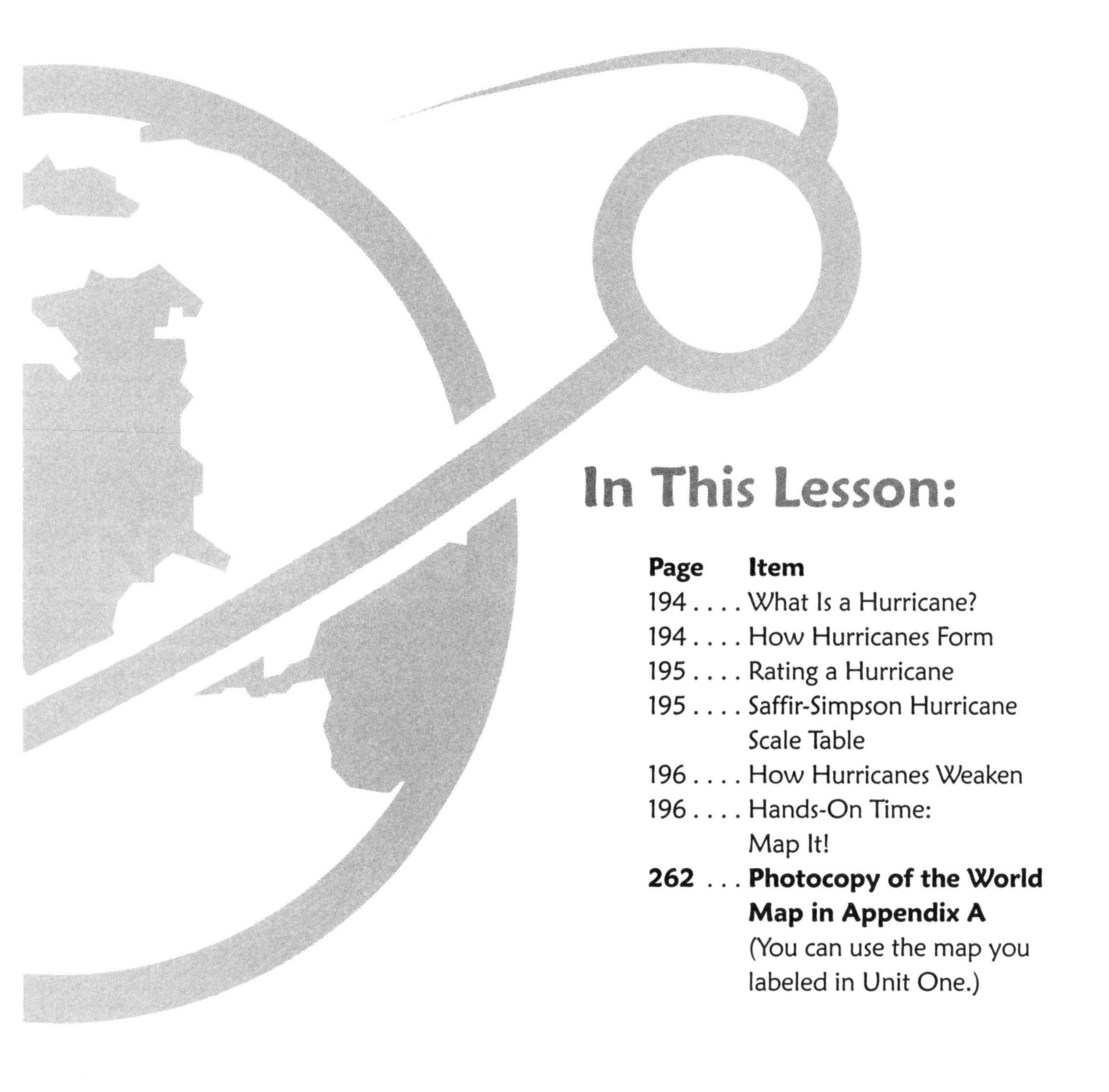

# In This Lesson:

# Lesson 20

# HURRICANES

## Teaching Time:

## In the Eye of the Storm

As we begin today, let's think about our weather lessons so far. We've explored seasons, clouds, precipitation, and violent weather such as thunderstorms and tornadoes. I've reminded you that God is in control of everything. In fact, He is Lord, or Master, over all our Earth. He created our Earth and everything in it. I like to remember this fact when I study, and especially, when I experience violent weather. It helps keep me calm. Today we're going to study one more type of violent weather . . . hurricanes.

**Scripture**

"Now when neither sun nor stars appeared for many days, and no small tempest beat on us, all hope that we would be saved was finally given up." (Acts 27:20)

Paul and his companions were out to sea when this took place. Read all of Acts 27 to see the whole story of a great storm at sea.

### Weather Patterns

Thunderstorms and tornadoes are more likely to occur in spring and summer than any other time of year. Did you know that? After studying the weather on Earth for many years, scientists have noticed a pattern. What's more, there are specific

**Name It!**

**hurricane**
Violent tropical storm that moves in a spiraling motion and has sustained winds of 74 miles per hour or greater.

**cyclone**
Same as a hurricane and a typhoon but located in a different part of the world.

**typhoon**
Same as a hurricane and cyclone but occurring in a different part of the world.

**tropical cyclones**
Common term for hurricanes, cyclones, and typhoons.

reasons for this pattern. For instance, these big storms happen more in spring and summer because air temperatures are heating up near the ground, the humidity is often high, and cold fronts are colliding with warm fronts. All together, these conditions create just the right situation for storms.

Like the violent weather we already mentioned, hurricanes follow patterns. Hurricane season runs from June to November. Most hurricanes happen in August, September, and October. As I write this lesson, it is August and there have already been eight major storms in the Atlantic Ocean this month. That is more than normal.

## What Is a Hurricane?

A **hurricane** is a violent tropical storm (usually in the warm, tropical waters of the Atlantic Ocean or Gulf of Mexico) that moves in a spiraling motion and has sustained winds of 74 miles per hour or greater. Wow! Did you get that? I'll explain. These are storms that usually begin near the coast of Africa. They move along propelled by the winds over the ocean, gaining strength and traveling west toward North America. As long as the winds near the surface of the ocean and those in the atmosphere continue to blow in the same direction and the storm is over warm waters, the storm can grow stronger.

You might have heard the terms **cyclone** and **typhoon** before. These are storms that have the same characteristics as hurricanes. The difference is where they are located. They can all be referred to as **tropical cyclones**. However, in the United States, Mexico, and the islands of the Atlantic Ocean, tropical cyclones are called hurricanes.

## How Hurricanes Form

Tropical cyclones such as hurricanes usually begin as a series of thunderstorms. Due to the strength of the winds and the wind patterns, the storms organize into a group, becoming a **tropical**

**depression**. If the conditions do not change very much, as far as winds and water temperature, the tropical depression can continue to grow and gain strength, becoming a **tropical storm**. As the storm gets more organized, it spins into a tighter circle, forming an **eye**. The eye is the center of the storm; it looks somewhat like the hole in a doughnut. This storm becomes a hurricane when the winds along the eye stay at 74 miles per hour or greater.

## Rating a Hurricane

Meteorologists rate hurricanes according to their wind speeds. The higher the winds, the higher the category number and the more severe the hurricane. The table included here will help you understand this concept. The scale used for this rating process is called the **Saffir-Simpson scale**, and it was developed in the 1970s.

### Saffir-Simpson Hurricane Scale

| Category | Sustained Winds (MPH) | Damage |
|---|---|---|
| 1 | 74–95 | Minimal |
| 2 | 96–110 | Moderate |
| 3 | 111–130 | Extensive |
| 4 | 131–155 | Extreme |
| 5 | >155 | Catastrophic |

Airplanes are used to fly into the eye and measure the wind speeds at the **wall**, or edge, of the eye. This is where winds are the greatest. Higher wind speeds create a smaller eye. The wall of the hurricane is the strongest part of the storm. However, inside the eye, the weather is calm and quite pleasing. Watch out, though. There's still another side of the wall that must pass, along with the rest of the storm before the danger is over.

You can learn about naming hurricanes at FEMA for Kids: www.fema.gov/kids/hunames.htm

**Name It!**

**tropical depression**
Storms organized into a group through strong winds and wind patterns.

**tropical storm**
A tropical depression that has continued to grow and gain strength.

**eye**
Tight circle located in the center of an organized storm.

**Saffir-Simpson scale**
Scale used to rate the strength of a hurricane.

**wall**
The edge of a hurricane's eye.

**Scripture**

Matthew 8:26 shows us that God controls the winds and the seas: "But He said to them, 'Why are you fearful, O you of little faith?' Then He arose and rebuked the winds and the sea, and there was a great calm."

## How Hurricanes Weaken

Thankfully, the storms once begun eventually come to an end. Many of these tropical storms never have much of an impact because they encounter cooler water and winds from other directions. These cause the storm to weaken before it ever reaches land.

However, sometimes tropical cyclones actually make landfall, bringing floodwaters and damaging winds. The good news is that the storms often die down soon after passing over land. With no more warm waters to feed the storm, and with changes in the wind patterns, there simply are not the ingredients present to give the storm the energy it needs to continue. Soon after landfall, a hurricane will lose strength and be downgraded to a tropical storm again and then down to a tropical depression. By then, it has usually moved back out to sea over cooler waters, and then it is gone.

In all of this, remember: There's no storm that God cannot calm. When I see a hurricane or typhoon on television that may cause much damage, I pray for the people who will be affected.

## Hands-On Time:
# Map It!

## Featured Activity

Earlier in this lesson, you learned that hurricanes are tropical cyclones. You also learned that these storms have different names depending on where in the world they occur. Today we're going to map out hurricanes, cyclones, and typhoons, which are all types of tropical cyclones.

*Note*: This is a technical and advanced effort. However, it can benefit all ages. To simplify, the teacher can label the map, longitude and latitude, first and then have the students join in with marking the territories for cyclones.

Additional Notes

## Objective

To learn more about tropical cyclones worldwide.

## Materials

- Photocopy of the world map in Appendix A (You can use the map you labeled in Unit One.)
- Colored pencils
- Globe or atlas

## Instructions

1. If using a blank map or using an atlas, find and label the oceans and continents.

2. Using your atlas as a guide, label the lines of longitude and the lines of latitude on your map. These help us pinpoint locations. Lines of longitude are the vertical lines on the map; lines of latitude are the horizontal lines.

3. Find the 180° line of longitude in your atlas. You will want to label the 180° line of longitude as the International Dateline. The dateline is not perfectly straight. It has zigs and zags, so watch out for this. (Hint: Look in the Pacific Ocean.) This line is important in understanding tropical cyclones around the world.

4. Using a colored pencil of your choice, lightly shade the Atlantic Ocean, the Gulf of Mexico, and the Caribbean Sea all one color.

5. Create a key, or guide, on the border of your map. Draw a small square using the colored pencil. Beside it write "Hurricanes."

6. Locate the Pacific Ocean. Shade the North Pacific Ocean east of the dateline, from Mexico up, with the *same color* as

Additional Notes

before. *Also*, using the same color, shade the South Pacific Ocean east of the 160° line of longitude up to the North Pacific Ocean.

7. Now, change colors. You're going to color the region where tropical cyclones are called "Typhoons." Make a square on your key with the color you are now going to use and label it accordingly.

8. Shade the northwest Pacific Ocean, west of the dateline, with this color.

9. Change colors again and mark your key. Label this "Cyclones."

10. Shade the remaining ocean waters in the Southern Hemisphere with this color.

11. Most, if not all, of the waters on your map are now colored. However, the more northern and southern parts of our world do not have tropical cyclones. The water just isn't warm enough. (Remember that the word *tropical* refers to warm.) However, I have learned that there are large polar storms called polar cyclones in these areas.

12. Add this map to your science notebook.

# Unit Five Wrap-Up

Now that you've completed the unit, it's time to wrap up all that you've learned and show what you know!

1. Copy this page and place a check mark in each box for work you've completed.
   - ❑ Lessons read
   - ❑ Daily Reading Sheets
   - ❑ Hands-On Time
   - ❑ Checking It Out forms
   - ❑ Vocabulary
   - ❑ Coloring Page

2. Now that you've accounted for the work up to this point, you should review all that you've studied. Here are some ways to do that.
   - ❑ Create a folderbook that covers the main ideas from each lesson. Instructions can be found in Appendix F.
   - ❑ Write a composition that reviews each lesson. See "Write About It!"—the composition worksheet and form in Appendix A.
     (*Note: If your science notebook is complete and you've done a good job with your Daily Reading Sheets, it should be fairly easy to get the facts you need for these summaries.*)

3. Show What You Know!—Complete the quiz on the next page to the best of your ability. After your teacher has reviewed your work, go back and use your book to answer anything you didn't know the first time through.

# Show What You Know!

Answer as many questions as you can without using your book or notes. You get **10,000** points for each correct response. After going through the quiz once with your book closed, have your teacher review your work. Then, open your book and try again. You get **5,000** points for each additional correct answer. So, **show what you know**!

1. The study of weather is called ____________________

2. Clouds are formed when water is evaporated and then ____________________, returns to liquid form, in the air.

3. Listed below are some major categories of clouds. They are Latin words. Fill in the blanks with their English translation.

   Cumulus ____________________

   Stratus ____________________

   Cirrus ____________________

   Nimbus ____________________

4. What is precipitation?

   ____________________

   ____________________

5. List two major ingredients for thunderstorms.

   ____________________

   ____________________

6. Funnel-cloud formations that develop within thunderstorms and reach the ground, often causing damage to trees and buildings, are called

   ____tornados____________________________________.

7. Hurricanes are common in (circle the answer) tropical waters / cool northern waters.

8. Hurricanes and typhoons are also known as

   ____tropical cyclones____________________________

9. The center of a hurricane is called the ____eye____________________

   and it is surrounded by the ____wall____________________.

10. Cool waters cause hurricanes to ____weaken____________________.

First Attempt ____________
*(number of correct responses X 10,000)*

Second Attempt + ____________
*(number of correct responses X 5,000)*

**Total Number of Points** = ____________

Rain
Change
Fair
Very Dry

# Unit Six
# Beyond Earth

## Unit Timeline

270 B.C. Aristarchus of Samos estimates the size of the Sun and its distance from Earth. He proposes that Earth goes around the Sun.

135 B.C. Hipparchus estimates the size of the Moon.

140 Claudius Ptolemaeus, known as Ptolemy, proposes his world system, which has the planets, Sun, and Moon moving around Earth. This Ptolemaic theory is a version of the geocentric (Earth-centered) theory.

1543 Nicolaus Copernicus publishes his theory of the solar system, which is a heliocentric (Sun-centered) theory. This model has been altered through time but is essentially the same.

1609 Galileo Galilei builds the first astronomical telescope and sees craters on the Moon and Jupiter's satellites. He also notices that Venus has moon-like phases.

1655 Christian Huygens develops a powerful telescope and identifies Saturn's rings.

1869 Norman Lokyer identifies a new element—later named helium—on the Sun.

*Timeline continues on next page.*

1899 Robert Goddard decides (upon climbing a cherry tree) to pursue his dream of space flight.

1926 Robert Goddard launches his first liquid-fuel rocket.

1958 The National Aeronautics and Space Administration (NASA) is established by President Dwight D. Eisenhower.

1959 The far side of Earth's moon is photographed.

1969 On July 20, Apollo 11 astronauts land on the Moon.

## Unit Six Vocabulary

- asteroid belt
- asteroids
- chromosphere
- comets
- craters
- ellipse
- first quarter moon
- full moon
- galaxies
- gas giants
- gravity
- helium
- hydrogen
- last quarter moon
- maria
- meteorites
- meteors
- Milky Way Galaxy
- new moon
- orbit (verb)
- photosphere
- rocky planets
- satellite
- solar corona
- solar eclipse
- solar system
- star
- Sun
- sunspots
- waning
- waning crescent moon
- waning gibbous moon
- waxing
- waxing crescent moon
- waxing gibbous moon

## Materials Needed for This Unit

- Science notebook
- Photocopy of the "Checking It Out" form in Appendix A
- Photocopy of the "Lunar Observation Chart" in Appendix A
- Sun-sensitive paper
- Dark stencils (and/or photocopy of "Sun-Print Images" page in Appendix A)
- Piece of glass, such as from a picture frame
- Dish large enough to accommodate the sun-sensitive paper
- Encyclopedia and recent magazine articles
- 10 paper plates
- Crayons and markers
- Index cards
- Globe
- Marker
- Tape or "sticky tack" removable adhesive
- Materials for Unit Wrap-Up (See page 243.)

NASA
SA

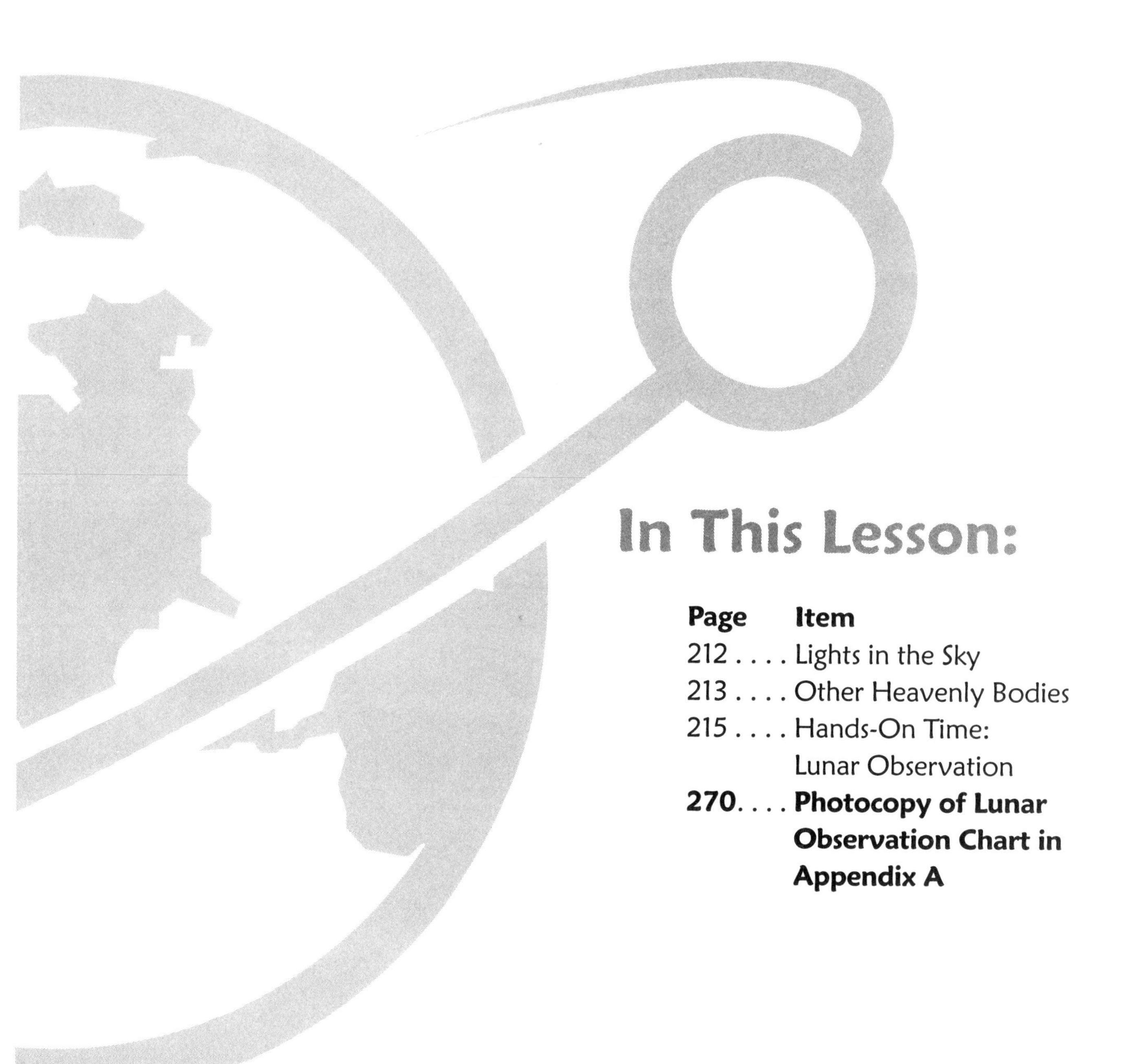

# In This Lesson:

Lesson 21

# MAKEUP OF THE UNIVERSE

Teaching Time:

## What's in Space?

Wow . . . the very idea of the universe is absolutely amazing! Sometimes when it is dark outside, I look up at the sky. If the night sky is clear, I usually spot the Moon first. As I look, I might see one or two lights in the sky, which I presume to be stars. But then I look a little closer, and do you know what I see? I see many, many lights. Now, I don't always know for sure if I'm seeing stars or planets, but still I am amazed. There is so very much beyond our Earth, I simply cannot comprehend it. There is an old hymn that says it better than I can. Read the words to the first verse to see what I mean.

 Scripture

"The LORD by wisdom founded the earth; by understanding He established the heavens . . . " (Proverbs 3:19)

**How Great Thou Art**
O Lord My God,
When I in awesome wonder
Consider all
The works Thy Hand hath made,
I see the stars,

Additional Notes

I hear the mighty thunder,
Thy pow'r throughout
The universe displayed;
Then sings my soul,
My Saviour God, to Thee,
How great Thou art!
How great Thou art!
Then sings my soul,
My Saviour God, to Thee,
How great Thou art!
How great Thou art!

*Originally a Swedish folk melody,*
"O Store Gud," *by Carl G. Boberg (1859–1940);*
*translated by Stuart K. Hine in 1899.*

Like the author of this hymn, I am moved to worship God's great power and majesty.

In this unit we are going to take a trip to outer space. We are going to learn a lot about the things we can see and about some that we can't see (without telescopes, that is.)

## Lights in the Sky

Here are some things I want you to think about as we get started. First of all, when you look at the sky in the daytime, what do you see? Most likely, you see blue sky, some clouds, and the Sun. What do you see when you look into the night sky? Now, can you tell me why you cannot see the stars in the daytime? We'll talk about it again, but right now I just want you to think about it.

The sky is full of many objects. First of all, there is the Sun, that wonderful ball of light we enjoy in the daytime. We will spend a whole lesson simply studying the Sun. Next, there is the Moon. One of the most beautiful things to see is the big, full Moon shining in the night sky. In all my life, I don't think I've ever heard anyone complain about the Moon! We will spend another lesson on the Moon. As you may or may not know, the sky also has planets. Some of these planets have their own moons. We will learn something about each of the planets in our sky in one of our lessons.

## Other Heavenly Bodies

Do you know how God has put our universe together? I confess that I only know a small bit about it all, but I love learning more! The universe is a vast expanse that goes farther than any of us can imagine. The distance to our Sun is over 93 million miles, and that's the closest star to us! How far away do you think the stars that look tiny are? In the vast expanse of the universe, there are groups of stars and planets that form **galaxies**. Earth is located in the **Milky Way Galaxy**. It is part of a group of planets and other heavenly bodies that revolve around the Sun. This group of planets is called our **solar system**. Our solar system has nine planets in it that we know of. However, as more sophisticated equipment is designed, more and more can be seen, and so scientists may still discover more planets in our solar system.

In addition to the planets and stars, there are **meteors** (**mee** tee ors) in our solar system. Meteors are collections of dust that move through the sky so fast they burn. Quickly, the dust is burned up and the fire goes out. Meteors are also known as shooting stars. Some meteors are large space rocks that race toward Earth and make it through our atmosphere. If they are not burned up in the atmosphere, they can land on Earth. These are called **meteorites**. When they hit Earth, they can form bowl-shaped depressions called **craters**. Some craters are only a few feet wide, while some are many miles wide. (This happens very rarely, so you don't have to be afraid.)

Comets and asteroids are two other things that can be found in our sky. **Comets** are large bodies of ice and dust. The main part of a comet is a gas that shines and causes the rest of the

**galaxies**
Large groupings of stars and planets found throughout the universe.

**Milky Way Galaxy**
The galaxy in which Earth is located.

**solar system**
A group of planets and other heavenly bodies that revolve around the Sun.

**meteors**
Collections of dust that move through the sky so fast they burn.

**meteorites**
Meteors that did not burn up as they fell through our atmosphere and thus hit Earth.

**craters**
Bowl-shaped depressions formed where meteorites hit Earth.

**comets**
Large bodies of ice and dust that form in the sky.

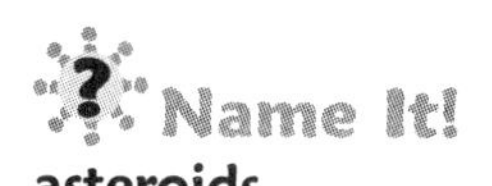

**asteroids**
Very large and rocky bodies that orbit the sun; sometimes called minor planets.

**asteroid belt**
A ring between Mars and Jupiter that contains almost all the asteroids in our solar system.

comet to shine. Comets have a comma shape with a head and a long tail. The path of some comets brings them near Earth where we can see them on a regular basis. When a comet is passing near Earth, it makes the news and people all over the world watch the sky.

**Asteroids** are sometimes called minor planets. They are very large, rocky bodies in the sky that orbit, or revolve around, the Sun. Between two of our planets, Mars and Jupiter, is very large ring called the **asteroid belt**. This belt contains almost all the asteroids in our solar system.

God has created a special place for us called Earth. Earth is part of a group of planets that form our solar system. Our solar system is part of the Milky Way Galaxy. Our Galaxy combines with other galaxies to form our magnificent universe.

Then sings my soul,
My Saviour God, to Thee.
How great Thou art!
How great Thou art!

Additional Notes

# Hands-On Time:
# Lunar Observation

## Featured Activity

*Note*: This will require a daily effort for most of Unit Six.

### Objective

To observe the Moon in several phases in preparation for the lesson on the Moon.

### Materials

- Photocopy of the "Lunar Observation Chart" in Appendix A
- Pencil
- Evening sky

### Instructions

1. Today you will begin observing the sky on a regular basis. Missing a night here or there will not ruin this exercise, so don't give up. You will need to choose a time when you can consistently observe the Moon. (Just before bedtime, soon after sunset, right after dinner, etc.) On your copy of the "Lunar Observation Chart," you will shade in the Moon each night as accurately as you can.

2. Always note the time that you go out. If you have to change the time of your observation for any reason, this will help you remember that information.

3. Also, record the date.

Additional Notes

4. The portion of the chart where you are to label the phase of the Moon will not be completed until you have your lesson on the Moon, so just file the form in your science notebook until then.

5. Have fun!

### One Step Further

For a little more fun, you can look at the stars in the sky.

- Use a book with constellation patterns to see if you can recognize any of the constellations.

- Also, look for steady lights in the night sky. These are usually planets. You can use binoculars to look a little closer. Remember, stars twinkle, while planets are steady lights. Airplanes, on the other hand, are obviously moving and are on a steady course. Man-made satellites move on a steady course, too, but they are much higher than airplanes and move faster.

- Visit a science center with a planetarium. There are several around the country that teach about the night sky.

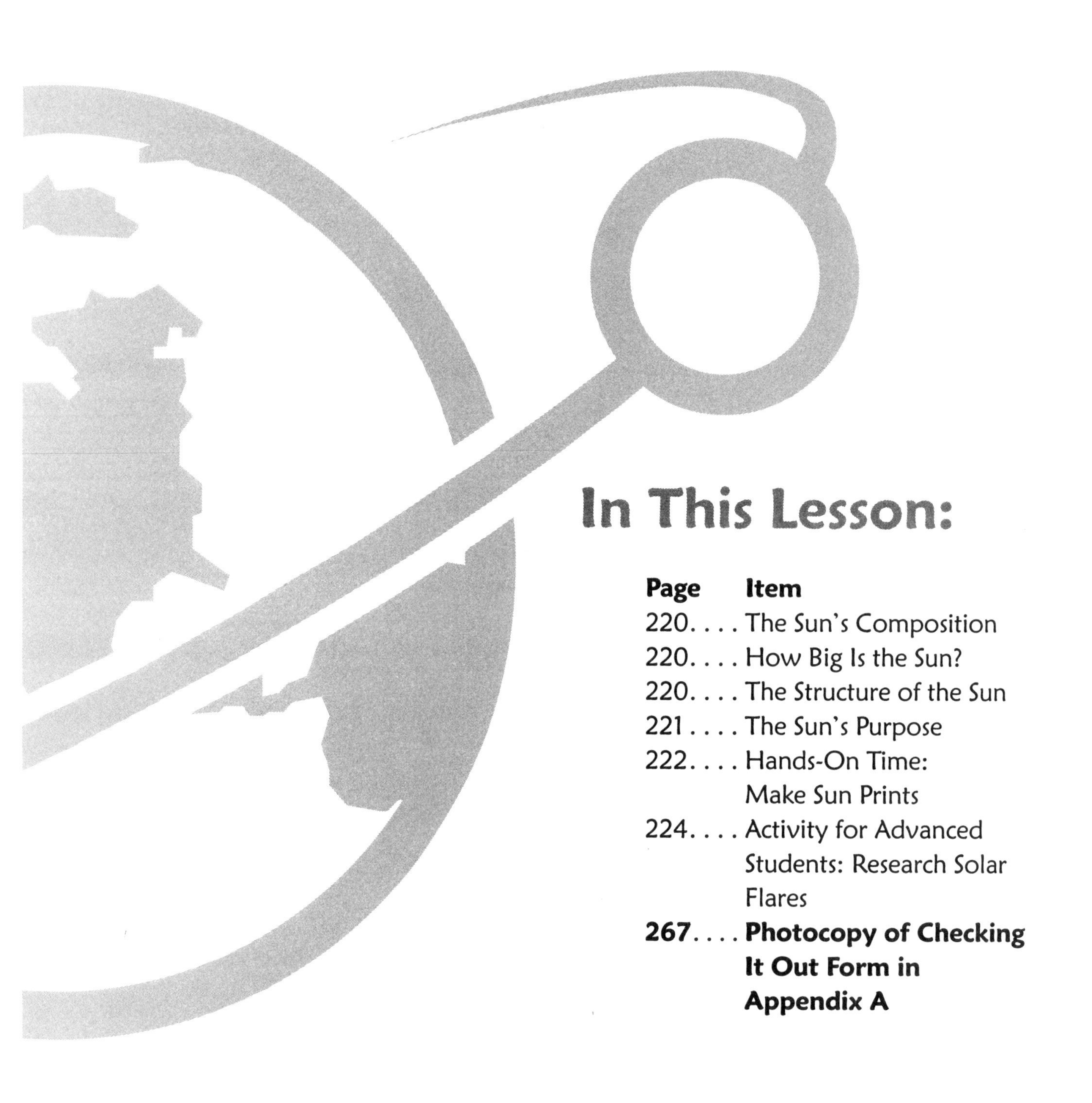

# In This Lesson:

# Lesson 22

# THE SUN

## Teaching Time:

## Solar Studies

Do you know what the Greek word *helios*, the Latin word *sol*, and the English word *sun* all have in common? If you said that they all mean the same thing, you're right. They are all words that represent our Sun. You know that the Sun warms you and provides light for our planet. But do you know what the Sun is made of? Do you know what its structure is? Can you imagine what Earth would be like without it? We're going to take a look at all of that and more in today's lesson. It is one more way to see how fabulous and complex our universe is.

The **Sun** is a star. Does that surprise you? I can understand if it does because from where we are, the Sun seems very different from the stars. First of all, we see the Sun in the daylight, and we see the stars at night. Second, the Sun is bright and hot. Stars are mere twinkles, and we feel no heat from them. Well, as much as that is true, it's not the whole truth. The Sun actually is a star. The reason that it seems so different is that it is much closer to us than other stars. In our first unit, you learned that

 **Scripture**

"He counts the number of the stars; He calls them all by name." (Psalm 147:4)

**Sun**

A star located closer to Earth than other stars. An enormous gaseous object in our solar system around which Earth and other planets revolve.

 **Warning!**

NEVER stare at the Sun, not even briefly, and NEVER stare at the Sun during an eclipse. The rays from the Sun are very powerful and will cause serious vision damage, possibly leading to blindness!

 **Name It!**

**star**
A heavenly body that is made of gas and produces its own light.

**hydrogen**
A gas that makes up about 75 percent of the Sun.

**helium**
A gas that makes up about 25 percent of the Sun.

**solar eclipse**
Occurs when the Moon's position blocks our view of the Sun.

daylight occurs because a side of Earth is facing the Sun. When that happens, the brightness and nearness of the Sun keep us from seeing all the other stars. They haven't disappeared; they're just too dim to compete with the Sun's brightness.

## The Sun's Composition

A **star** is a heavenly body that is made of gas and produces its own light. The Sun is made of the gases **hydrogen** (**high** dro jen) and **helium** (**heel** ee um). It is about 75 percent hydrogen and about 25 percent helium. In addition, a very small percentage of the Sun is metals. These gases burn and produce incredible amounts of heat. The light we see is the burning of the gases on the Sun. That is also the heat we feel. Like other stars, however, it does not have a solid base. Earth is solid; the Sun is gaseous.

## How Big Is the Sun?

The Sun is the largest object in our solar system. In fact, nothing else can begin to compete. It would take about 100 Earths to span the width of the Sun. From our view, the Moon may look as if it's the same size as the Sun, but the truth is that the Moon is much smaller. It looks the same size because it is much nearer to Earth than the Sun is. Sometimes the Moon's position blocks our view of the Sun. This is called a **solar eclipse** (**so** lar ee **clips**).

The Sun is approximately 93 million miles away from Earth. The Sun is near Earth in terms of the size of the universe and compared to other stars. It does not move, even though it may seem that way to those of us on Earth. Earth revolves around the Sun, as do all of the other planets in our solar system.

## The Structure of the Sun

Now that you know that the Sun is the enormous gaseous object in our solar system around which Earth and the other planets revolve, let's take a look at the structure of the Sun. The Sun has some distinct areas that scientists have named and now study. The first is the **photosphere** (**fo** tus fear). The photosphere is the surface of the Sun. The next major area is the **solar corona** (**so** lar ka **row** na). The solar corona is the outer atmosphere of the Sun, and it can be seen during an eclipse. In between the photosphere and the solar corona is the **chromosphere** (**kro** mus fear). The Sun also has features called **sunspots**. Sunspots are areas where the surface is cooler than the rest of the surface.

CAUTION: Be sure to remember and heed the warning issued earlier in this lesson—NEVER stare at the Sun, not even briefly, and NEVER stare at the Sun during an eclipse!

## The Sun's Purpose

Do you know that we could not live without the Sun? It's necessary for all life on Earth. When God created the universe, He put the Sun in exactly the right position for life to exist on Earth. If it were any closer, we'd burn. If it were any farther away, we'd freeze. It makes plants and people thrive and grow. It warms us. It provides daylight for us. Can you imagine having no sunlight at all . . . ever? I can't. I'm glad God is in charge! I'm thankful for the way He provides for us. Our Sun is vital to life on Earth.

### Discovery Zone

To find out more about the structure of the Sun, go to this fascinating Web site: http://sunearth.gsfc.nasa.gov. Choose the multimedia viewer button (third picture on the left under "Recommended Viewing"). Then choose "Illustrations" and see, especially, illustrations 01, "The Solar Interior," and 02, "Seeing Inside the Sun," for good views of the Sun's structure.

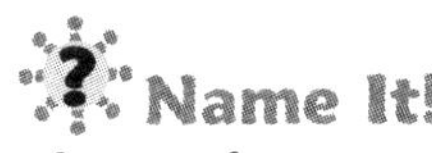


**photosphere**
Surface of the Sun.

**solar corona**
Outer atmosphere of the Sun; can be seen during an eclipse.

**chromosphere**
Layer of the Sun's atmosphere located between the photosphere and the solar corona.

**sunspots**
Areas where the surface of the Sun is cooler than the rest of its surface.

Additional Notes

# Hands-On Time: Make Sun Prints

## Featured Activity

### Objective

To see the effects of the Sun on Earth and consider the value of protection from the Sun.

### Materials

- Sun-sensitive paper
  *Note*: Before you remove the paper from its packaging, you will need to be in a dimly lit room. Otherwise your paper may begin to change color before you're ready.

- Dark stencils (Use cutouts of shapes, cardboard designs, flat leaves, keys, etc.—anything that will prevent the light from passing through. A page of sun-print images has been provided in Appendix A.)

- Piece of glass (One out of a picture frame would work well, with parent's permission.)

- Dish large enough to accommodate your paper, filled with water.

- Photocopy of the "Checking It Out" form in Appendix A

- Encyclopedia and recent magazine articles

## Instructions

Additional Notes

1. Place paper on a firm, flat surface, such as a cookie sheet.
2. Place desired stencils on the paper.
3. Cover the paper and stencils with the piece of glass (or heavy clear plastic) to keep the objects from moving around or blowing away.
4. Place in sunlight until the dark paper turns a very light blue. This should take less than five minutes.
5. Keeping your paper out of the light, remove all the objects on the paper and place the paper in a dish of water for one minute.
6. Lay flat to dry. As the paper dries, the images will become more clearly defined.
7. Answer the following:
   - What apparently caused the paper to change colors?
   - What difference did the stencils make to the overall appearance of the paper?
   - What can you learn about sunlight from this demonstration?
   - Explain how you think this could relate to protecting yourself from the Sun.
8. Complete a "Checking It Out" form and place it in your science notebook; include your responses to the above questions.

Additional Notes

# Activity for Advanced Students: Research Solar Flares

## Objective

To learn more about the Sun and its activity.

## Instructions

1. Using the Internet as well as recent magazine articles and an encyclopedia, research the Sun and, more specifically, solar flares.

2. Write a report detailing what you learn. If possible, include photographs from the Internet or magazine. (Remember to always tell where you got your information when using someone else's work.)

3. Add this report to your science notebook.

# In This Lesson:

# Lesson 23

# EARTH'S MOON

## Teaching Time:
## In Orbit

Once upon a time, when I was a little girl, I remember riding along in the car and watching the Moon following us. It went behind the trees, along the fields, and kept right up with our car. I distinctly remember asking my parents about it and having them explain to me that the Moon wasn't following us. Nonetheless, it looked that way to me!

Another specific memory I have is of the first time an astronaut walked on the Moon. I really couldn't believe that people had traveled to the Moon and were able to walk on it. That is still an amazing notion almost forty years later. I can tell you, though, that I lived most of my life without ever learning very much about the Moon. I don't know why, but I didn't. I've often wished that I had learned more about it when I was younger. I don't want you to be like me and miss out on studying this neat object in our sky. So, today we're going to make a lunar visit of sorts and see what we can learn.

**Scripture**
"He appointed the moon for seasons; the sun knows its going down." (Psalm 104:19)

**Moon's Elliptical Orbit of Earth**

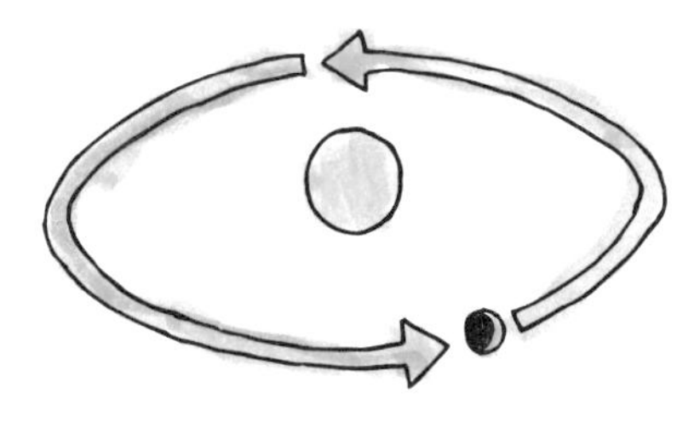

**Name It!**

**orbit**
As a verb: To revolve around.

**ellipse**
An oval shape.

**satellite**
An object that orbits another object.

## How Near Is the Moon to Earth?

Earth's Moon is the brightest object in our night sky. It is about 240,000 miles away from Earth. Sometimes it's actually closer and sometimes a little further away. This is because it **orbits**, or moves around, our Earth in an elliptical pattern, not a perfect circle. An **ellipse** (ee **lips**) is like an oval. See the accompanying illustration of the Moon's elliptical orbit.

## The Light of the Moon

As I mentioned, the Moon is the brightest object in our night sky. This could make you think that, like the Sun, the Moon produces its own light. However, you would be mistaken. The reason the Moon looks bright is because the Sun is shining on it. Isn't that interesting? It's like Earth's daylight. The light is caused by the Sun, not by Earth. In the same way, the light we see when we look at the Moon is caused by the Sun, not the Moon.

## The Moon Is a Satellite

Do you know that the Moon revolves and rotates? The word orbit that you just learned means to "move around" an object. That's the same thing as revolving. The Moon revolves around Earth. Do you remember what Earth revolves around? Yes, the Sun! An object that orbits another object is called a **satellite** (**sat** uh lite). We have man-made satellites in space, but the Moon is a God-made satellite.

The Moon also rotates on an axis. Now, you may remember that Earth rotates once on its axis every 24 hours, and it takes a full year to revolve around the Sun. The Moon, on the other hand, rotates at about the same rate that it revolves around Earth. That means that the same side of the Moon is always toward Earth. It takes a little more than 27 days for the Moon to orbit Earth. You might sometimes hear people refer to the "far side" of the Moon. That is the side of the Moon that never faces Earth.

# The Waxing and Waning Moon

Have you ever wondered about the way the Moon appears to change shape in the sky? This was a puzzle for me for a long time. As Earth orbits the Sun and the Moon orbits Earth, the Moon's position in relation to Earth and the Sun is changing. Since the light of the Moon is caused by the Sun shining on it, and the Moon and Earth are in constant motion, our view is gradually changing. See the accompanying chart that illustrates the Moon's phases.

Did you know that the Moon has seas? Well, to explain, the Moon does not have bodies of water. From Earth, we see large areas of the Moon that are dark. Long ago, people on Earth speculated that these were seas like we have on Earth. They named these dark spots **maria**, which is plural for *mare*, meaning "sea." Modern science tells us, however, that these are low spots or plains, and they do not contain water.

The Moon is a dusty, rocky body. Some people believe that the Moon split off from Earth in a giant space explosion. However, the Bible discusses God putting the lights in the sky. The Moon is a satellite of Earth. Other planets have satellites as well. We will learn more about them in another lesson. By the way, you may think that the Moon is only out at night. That is not true. There are times when the Moon can be clearly seen in the daylight. You can learn a good deal just by observing the Moon for yourself. Have fun!

**new moon**
In this phase, the Moon appears totally black; we cannot see the sunlit portion.

**waxing crescent moon**
In this phase, the sunlit portion of the Moon can be seen becoming larger each night, as the Moon and Earth change positions in relation to the Sun.

**first quarter moon**
In this phase, the Moon looks like a half circle.

**waxing gibbous moon**
In this phase, the Moon appears to be a little larger than a perfect half circle.

**full moon**
In this phase, the entire side of the Moon is sunlit and visible as a full circle.

**waning**
In this phase of the Moon, which follows the full moon, the Moon appears to get smaller.

**waxing**
In this phase of the Moon, between the new moon and the full moon, the Moon appears to get larger.

**waning gibbous moon**
In this phase, which follows the full moon, the Moon is on its way to becoming a half circle again.

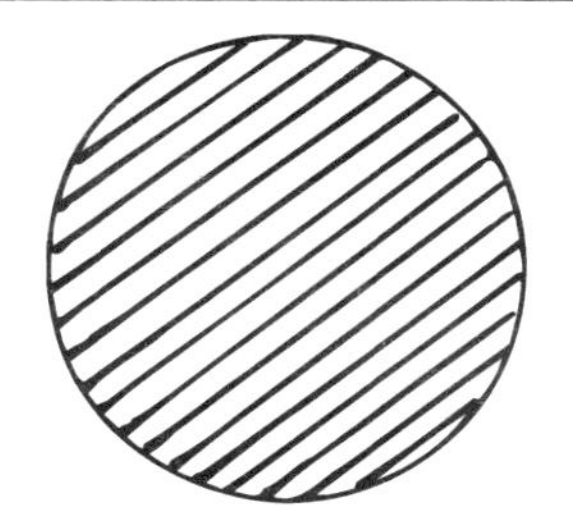

**New Moon**

The beginning of the cycle is called a **new moon**. A new moon looks totally black from Earth. We cannot see the sunlit portion at all.

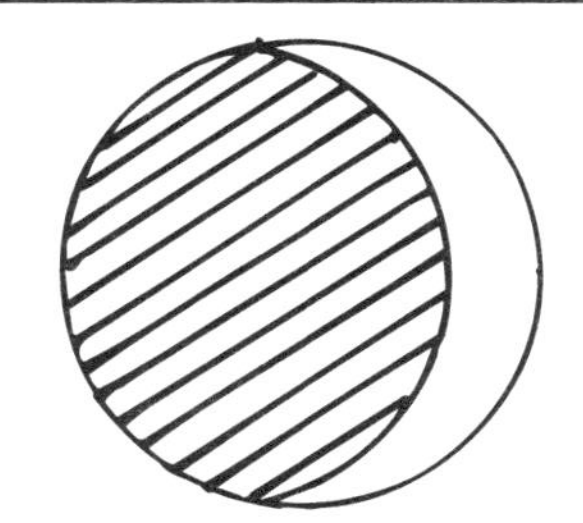

**Waxing Crescent Moon**

As the Moon and Earth change positions in relation to the Sun, we see a little more of the sunlit portion of the Moon each night. This is called a **waxing crescent moon**.

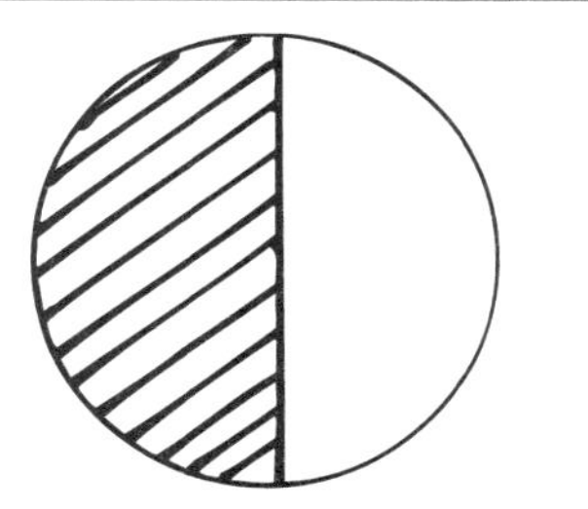

**First Quarter Moon**

Before long, we see a **first quarter moon**. A first quarter moon looks like a perfect half circle.

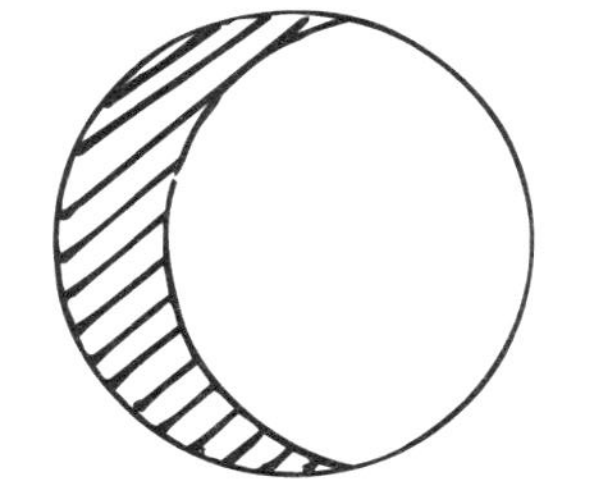

**Waxing Gibbous Moon**

When the Moon gets a little larger than a perfect half circle, it is known as a **waxing gibbous moon**.

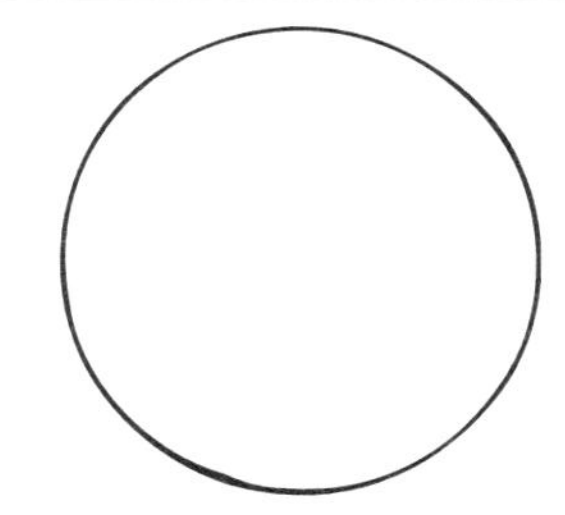

**Full Moon**

As the Moon continues its movement, the time comes when we can see the entire sunlit portion. This is called a **full moon**. This marks the halfway point.

**Waning Gibbous Moon**

From this point on, Earth's view of the sunlit portion of the Moon gets smaller, so the Moon looks smaller. In this stage, the Moon is **waning**; on its way to a half circle again, it is a **waning gibbous moon**.

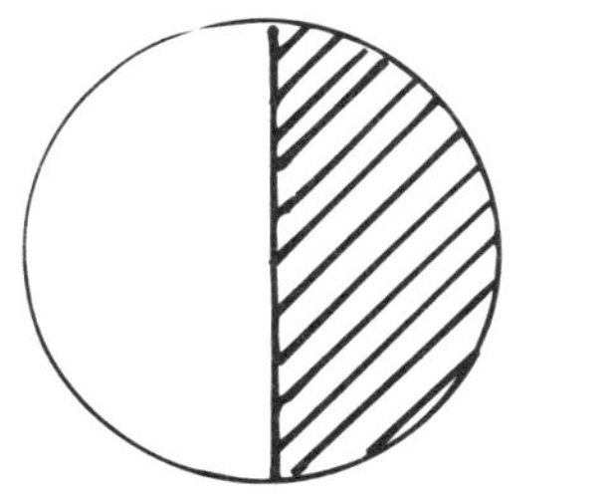

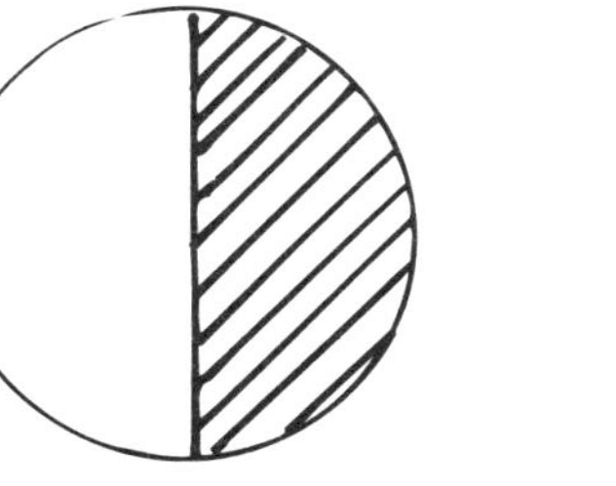

**Last Quarter Moon**

When we get to the half circle again, it is called a **last quarter moon**.

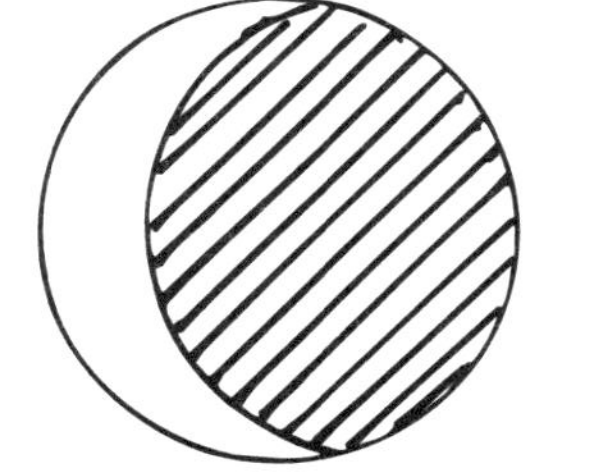

**Waning Crescent Moon**

The phase between a last quarter moon and a new moon is called a **waning crescent moon**. This looks like a sliver of a moon.

**New Moon**

Finally, the cycle is complete and we're back to a new moon.

## Review It!

*May be done orally or written.*

1. The Moon orbits Earth in an (oval) elliptical pattern.

2. What makes the Moon bright?

   the sun

3. What do we call an object that revolves around another object? satellite

4. It takes 27 days for the Moon to revolve around Earth.

5. How fast does the Moon rotate on its axis?

   27 days (same!)

6. When the Moon seems to get bigger, we say it is a waxing moon.

7. We say the Moon is waning when we see less of it each night.

### Name It!

**last quarter moon**
In this phase, which follows the full moon, the Moon appears to be a half circle.

**waning crescent moon**
In this phase, the Moon appears as a sliver, just prior to the new moon.

**maria**
Low spots or plains on the Moon that do not contain water but that are visible as large dark areas and were once thought to hold water. (*Maria* is the plural for *mare*, meaning "sea.")

Additional Notes

# Hands-On Time: Make a Lunar Wall Chart

## Featured Activity

### Objective

To help students understand and memorize the phases of the Moon and use this knowledge to complete their "Lunar Observation Chart," begun in Lesson 21.

### Materials

- 10 paper plates
- Crayons
- Index cards
- Globe
- Marker
- Tape or "sticky tack" removable adhesive

### Instructions

First, let's make a wall chart illustrating the phases of the Moon.

1. Use a black crayon to color one paper plate for each phase of the Moon, as illustrated in "The Phases of the Moon" chart in this lesson.

2. Using another paper plate, and a globe for reference, color the plate to look like a portion of Earth. You might want to choose the side that has the continent you live on.

3. Now, look again at the chart of the Moon's phases and use sticky tack to apply your plates to the wall in the same order, with your "Earth" in the center.

4. Label the index cards with the names of the phases. Adhere the cards to the wall beside their appropriate phases.

5. Now you have a "chart" of your own for easy studying. You can practice by removing the index cards from the wall, mixing them up and seeing if you can return them to their proper spots.

6. Take a picture for your science notebook!

Now, let's use this information to complete the "Lunar Observation Chart" you started in Lesson 21.

1. Look at the display on your wall and find the corresponding phases on your chart.

2. Then, label the phases on your chart. When the Moon is getting bigger, it is waxing. When the Moon is getting smaller, it is waning.

3. Try to label each box on your chart. Hopefully, you can see how each night the Moon got a little bigger or a little smaller depending on where it was in its phase.

4. If you continue your observation for at least 28 days, you will eventually get back to where the Moon was at the beginning of your observation.

**Additional Notes**

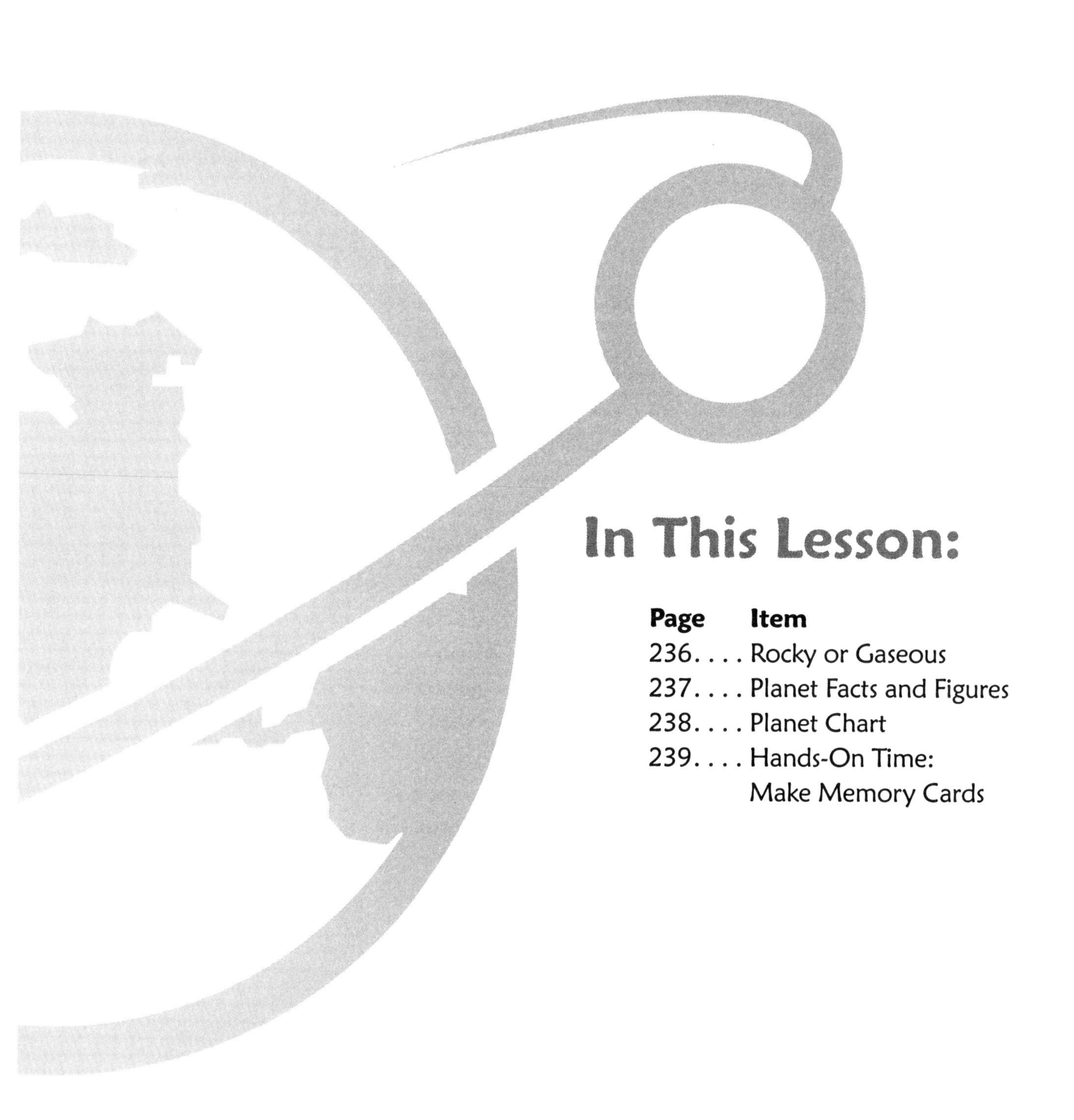

# In This Lesson:

Lesson 24

# THE EIGHT OTHER PLANETS

## Teaching Time:
## Out in Space

Our unit on space is winding down. So far you have learned that there are several types of bodies in space. There are stars, planets, and asteroids. There are also comets. You've learned that the universe God created has galaxies and that Earth is located in a solar system that is in the Milky Way Galaxy. You now know, if you didn't before, that Earth is a planet that revolves around the Sun, which is a star. You also know that Earth has one God-made satellite and that is the Moon. Whoosh! It might seem like that's enough to know, but then we'd miss out. In fact, we've spent this whole book covering only the planet Earth and a few outer-space elements. There are eight other planets within our solar system! We can't stop now!

**Scripture**
"O LORD, our Lord, how excellent is Your name in all the earth, Who have set Your glory above the heavens!" (Psalm 8:1)

### Eight More Planets by Name

All right then, let's get started. Earth is one of nine planets that revolve around the Sun. It is the third planet from the Sun. Can

 **Discovery Zone**
Check out http://hubblesite.org. Choose "gallery" to see photos taken of our solar system. There are many other fun things to do at this Web site!

**Name It!**

**rocky planets**
Planets composed mostly of rock and metal: Mercury, Venus, Earth, and Mars.

**gas giants**
Planets composed mostly of gas: Jupiter, Saturn, Uranus, and Neptune.

you name any of the other planets? Did you think of Mars? Yes, Mars is one of the nine planets in our solar system. Let me list them all for you, in order from the planet closest to the Sun to the planet farthest from the Sun:

1. Mercury
2. Venus
3. Earth
4. Mars
5. Jupiter
6. Saturn
7. Uranus
8. Neptune
9. Pluto

*Note*: Neptune and Pluto cross each other as they orbit the Sun, so sometimes Pluto is eighth from the Sun and Neptune is ninth.

## Rocky or Gaseous

Most of the planets around the Sun can be grouped, or classified, in one of two ways. First of all, there are the **rocky planets**. Earth is an example of a rocky planet. The interiors of rocky planets are believed to consist of rock and metal. There are four rocky planets: Mercury, Venus, Earth, and Mars. Notice that all these planets are the ones closest to the Sun.

The next four planets from the Sun—Jupiter, Saturn, Uranus, and Neptune— belong to a group of planets called the **gas giants**. As you might guess, these are composed mostly of gas. There may be metal cores in some cases and possibly rocky cores in others. However, they are still largely gas.

Pluto is not in either category. It seems to be both rocky and gaseous.

## Planet Facts and Figures

Let's look at the "Planet Chart" in this lesson to see more about the planets. Review the chart. Did you learn anything new? Did you notice that some planets have no moons, or satellites, while others have many? Novice, inexperienced sky watchers can actually see four of Jupiter's moons. Look at the gas planets. Do you see anything they have in common? I hope you noticed that they all have rings. Scientists have known about Saturn's rings for a long time, but they have only recently learned that the other gas giants also have rings.

You may be wondering how all the things in space stay in their proper place. To me, the most obvious answer is that God is an orderly God and a perfect designer. In His divine order He has also given us a scientific reason. It's called gravity. **Gravity** is a force caused by the mass of a body that attracts lesser masses to it. Earth has gravity and that is what keeps us from floating off into space. Jupiter has so many satellites because of its huge mass. In fact, Jupiter has more mass than any other planet.

Scientists are discovering more and more about our solar system all the time. The many advances in technology are making this possible. There's even a belief that there may be a 10th planet that is beyond Pluto. Scientists have admitted they don't know for sure how much is out in space that they know nothing about. Space, and all that is in space, teaches me about the limitless power of God. As you look into the night sky, you—like David—can consider how small man is in comparison to the universe. (See Psalm 8.)

### Helpful Hint

Mass is technically the measure of matter in a body. It is possible to have something that is large that has little mass. Do not get hung up on this terminology at this elementary stage. I simply desired to use the proper term. This is covered later at the high school level.

### Name It!

**gravity**

A force caused by the mass of a body that attracts lesser masses to it.

## Planet Chart

| Planet Facts | Mercury | Venus | Earth | Mars | Jupiter | Saturn | Uranus | Neptune | Pluto |
|---|---|---|---|---|---|---|---|---|---|
| **Rocky or Gas?** | Rocky | Rocky | Rocky | Rocky | Gas | Gas | Gas | Gas | Both |
| **Distance from Sun** | 36 million miles | 65 million miles | 93 million miles | 137 million miles | 466 million miles | 856 million miles | 1.8 billion miles | 2.8 billion miles | 3.6 billion miles |
| **Time to Orbit Sun** | 88 Earth days | 224 Earth days | 365¼ Earth days | 687 Earth days | 12 Earth years | 29½ Earth years | 84 Earth years | 165 Earth years | 248 Earth years |
| **Time to Rotate on Axis** | 59 Earth days | 243 Earth days | 24 Earth hours | 1 Earth day | 10 Earth hours | 10 Earth hours | 18 Earth hours | 19 Earth hours | 6 Earth days |
| **Size** *(1 to 9: 1 = lrgst 9 = smlst)* | 8 | 6 | 5 | 7 | 1 | 2 | 3 | 4 | 9 |
| **No. of Known Moons** | 0 | 6 | 1 | 2 | 61 | 30 | 21 | 11 | 1 |
| **Other Info** | Moves faster than any other planet. | Most visible in our night sky. | Only planet with known life. | Called the Red Planet. | Has a faint ring. Its Great Red Spot is about a 400-yr-old storm. | Has noticeable rings that look like one ring, but are actually many. | Has faint rings. | Has narrow rings. | Its orbit crosses Neptune's, so sometimes it is the 8th farthest planet from the Sun. |

Additional Notes

## Hands-On Time:
# Make Memory Cards

## Featured Activity

### Objective
To memorize basic planet facts.

### Materials
- Index cards
- Markers, pencil

### Instructions
1. Using the design here, or creating one of your own, make a card for each planet.
2. Use the cards daily for about five minutes to memorize some basic planet facts.
3. You can add other facts if you desire.

## Additional Notes

Side 1

# Planet Name

(You might like to color a picture of the planet or cut out a color print from your computer.)

Side 2

# Planet Name

Order from the Sun:

1st 2nd 3rd 4th 5th 6th 7th 8th 9th

Number of Moons:

Distance from the Sun:

Choose One: Rocky Gas Giant Both

# Unit Six Wrap-Up

Now that you've completed the unit, it's time to wrap up all that you've learned and show what you know!

1. Copy this page and place a check mark in each box for work you've completed.
   - ❑ Lessons read
   - ❑ Daily Reading Sheets
   - ❑ Hands-On Time
   - ❑ Checking It Out forms
   - ❑ Vocabulary
   - ❑ Coloring Page

2. Now that you've accounted for the work up to this point, you should review all that you've studied. Here are some ways to do that.
   - ❑ Create a folderbook that covers the main ideas from each lesson. Instructions can be found in Appendix F.
   - ❑ Write a composition that reviews each lesson. See "Write About It!"—the composition worksheet and form in Appendix A.
     (*Note: If your science notebook is complete and you've done a good job with your Daily Reading Sheets, it should be fairly easy to get the facts you need for these summaries.*)

3. Show What You Know!—Complete the quiz on the next page to the best of your ability. After your teacher has reviewed your work, go back and use your book to answer anything you didn't know the first time through.

# Show What You Know!

Answer as many questions as you can without using your book or notes. You get **10,000** points for each correct response. After going through the quiz once with your book closed, have your teacher review your work. Then, open your book and try again. You get **5,000** points for each additional correct answer. So, **show what you know**!

1. Name the galaxy that contains Earth.______________________________.
2. Earth is part of a group of planets that orbit the Sun. This is a ______________ system.
3. Collections of dust that move through our system so fast they burn and eventually burn up are ______________________________.
4. If these burning particles reach Earth, they are called ____________________.
5. ____________________are large bodies of ice and dust shaped like commas that move through the solar system on a regular path.
6. Asteroids are sometimes called ______________________________ due to their size and the fact that they also orbit the Sun.
7. Stars are made of gas and produce their own ______________________________.
8. The largest body in our solar system is the ______________________________.
9. When the Moon appears to get bigger each night, it is called a ______________ moon.
10. When the Moon appears to get smaller each night, it is called a ______________ moon.
11. True / False (circle one) The Moon revolves around Earth.

12. True / False (circle one) The Moon rotates on its axis.

13. How long does the Moon take to rotate once on its axis?

_______________________________________________

14. What is the smallest known planet? ______________________________

15. Earth is the (circle one) first / second / third planet from the Sun.

16. How many planets are there in our solar system? ____________________

17. Name as many of the planets as you can.

_______________________________________________

_______________________________________________

_______________________________________________

First Attempt ______________
*(number of correct responses X 10,000)*

Second Attempt + ______________
*(number of correct responses X 5,000)*

**Total Number of Points** = ______________

# GLOSSARY

**Active volcano** — An erupting volcano.

**Air composition** — The "ingredient list" of the air.

**Alpine glacier** — Glacier that forms in mountainous areas; also called a valley glacier.

**Antarctica** — The continent surrounding the South Pole, almost entirely covered by an ice sheet, making it a continental glacier and the world's largest glacier.

**Aquifer** — Layers of rocks, soil, and sand that contain water.

**Asteroid belt** — A ring between Mars and Jupiter that contains almost all the asteroids in our solar system.

**Asteroids** — Very large and rocky bodies that orbit the sun; sometimes called minor planets.

**Atmosphere** — The gases, necessary to support life, that surround our planet.

**Atmospheric pressure** — The pressure of the air all around us.

**Axis** — Imaginary rod through the center of Earth around which Earth rotates.

**Barometer** — Instrument that measures changes in air pressure.

**Calcite** — Mineral that results from carbonic acid mixing with dissolved limestone, deposited by water and responsible for many speleothems.

**Caldera** — Volcano formed low to the ground, caused by the ground's collapsing after an explosive eruption.

**Calving** — The breaking off and forming of icebergs from glaciers.

**Carbonic acid** — Acid formed by rainwater mixing with carbon dioxide in the air; responsible for forming caverns slowly.

**Caverns** — Caves formed by water and acid that slowly erode soluble rock.

**Chasms** — Very deep valleys or canyons, often found in glaciers.

**Chemical rock** — A type of sedimentary rock formed by the evaporation process.

**Chromosphere** — Layer of the Sun's atmosphere located between the photosphere and the solar corona.

**Cinder cone volcano** — Volcano caused by an explosive eruption, cone-shaped and relatively small in size.

**Cirrus** — Thin, white clouds containing ice crystals; found at 20,000 feet and higher; name translated means "curl of hair."

**Circumference** — Measurement around an object.

**Clastic rock** — A type of sedimentary rock formed in water.

**Clouds** — Collections of very small water droplets and ice crystals that form when water is evaporated from the ground and then returns to liquid form in the air.

**Cold front** — Cool air pushing in on warm air.

**Columns** — Cave formations formed when stalagmites and stalactites grow together.

**Comets** — Large bodies of ice and dust that form in the sky.

**Composite volcano** — Volcano caused by an explosive eruption, cone-shaped and large in size due to repeated eruptions over a long period of time.

**Composition** — One of two methods of studying the atmosphere; breaks the atmosphere into two layers, the homosphere and the heterosphere.

**Condense** — To change back to liquid form.

**Continental glacier** — Enormous ice sheet that covers most of a continent or large island.

**Convection** — Rising warm air.

**Coral caves** — Caves formed by coral colonies joining each other in shallow waters.

**Core** — The area inside Earth made up of the inner and outer cores combined.

**Crater** — Sunken area inside a volcano where magma collects.

**Craters** — Bowl-shaped depressions formed where meteorites hit Earth.

**Crust** — Outermost layer of Earth; about six miles thick.

**Cumulus** — Clouds formed by rapidly rising air, extending up to 39,000 feet and higher; name translated means "heap."

**Current** — A flow of water.

**Cyclone** — Same as a hurricane and a typhoon but located in a different part of the world.

**Dehydrated** — Serious medical condition in which your body does not have enough water to function properly.

**Deposit** — As a verb: To put, place, or leave something somewhere; in this case, by a natural process.

**Doppler radar** — Weather-predicting tool that allows the inside of a storm to be seen through computer images.

**Dormant volcano** — Volcano with long periods of inactivity.

**Drainage** — The natural or artificial removal of water from a given area.

**Dry snow** — Snow that occurs in very cold air when the snowflakes are lighter.

**Earthquakes** — Vibrations caused by the shifting of the tectonic plates of Earth's crust; often mild but sometimes very destructive.

**Ellipse** — An oval shape.

**Epicenter** — Area on the surface of Earth just above an earthquake's focus.

**Equator** — Imaginary line that separates the Northern Hemisphere from the Southern Hemisphere; labeled as the 0 degree line of latitude.

**Erosion** — Process of washing away.

**Eruption** — A reaction caused by pressure building up inside Earth, forcing magma up through a volcano's vent and out of the volcano.

**Evaporate** — To change from a liquid to a vapor.

**Extinct volcano** — Volcano that no longer erupts.

**Eye** — Tight circle located in the center of an organized storm.

**Faults** — Cracks in the tectonic plates; responsible for many earthquakes.

**Firmament** — An expanse; the sky and everything beyond Earth, including the Sun, Moon, and stars.

**Firn** — Layer of hard-packed snow on a glacier.

**First quarter moon** — In this phase, the Moon looks like a half circle.

**Focus** — Spot underground where the central power of an earthquake is located.

**Freezing** — 32 degrees Fahrenheit.

**Freezing rain** — Rain that falls from the clouds and freezes when it comes in contact with the ground or things on the ground. Occurs when the ground is below 32 degrees Fahrenheit, but the air temperature is higher.

**Freshwater** — All the water on Earth that's not saltwater; found in lakes, streams, rivers, and ponds, as well as in precipitation.

**Fujita scale** — Scale used to rate the strength of a tornado.

**Fujita, Theodore** — Developed the Fujita scale in 1971.

**Full moon** — In this phase, the entire side of the Moon is sunlit and visible as a full circle.

**Funnel cloud** — Spinning of the air and the water vapor inside.

**Galaxies** — Large groupings of stars and planets found throughout the universe.

**Gas giants** — Planets composed mostly of gas: Jupiter, Saturn, Uranus, and Neptune.

**Geologists** — Scientists who study rocks.

**Geology** — The study of Earth's rocks.

**Glacier** — Large mass of ice that moves slowly.

**Gravity** — A force caused by the mass of a body that attracts lesser masses to it.

**Groundwater** — Water that drains through the soil and is stored underground in aquifers.

**Hail** — Water droplets that have been tossed up and down in violent cumulonimbus clouds freezing to ice, collecting more water, freezing more, over and over until balls of ice are formed.

**Helium** — A gas that makes up about 25 percent of the Sun.

**Hetero** — Different.

**Heterosphere** — One of two layers that make up the composition of the atmosphere.

**Homos** — Same.

**Homosphere** — One of two layers that make up the composition of the atmosphere; composed of the troposphere, stratosphere. and mesosphere.

**Humidity** — Moisture content of the air.

**Hurricane** — Violent tropical storm that moves in a spiraling motion and has sustained winds of 74 miles per hour or greater.

**Hydrogen** — A gas that makes up about 75 percent of the Sun.

**Hydrologic** cycle — Process by which water moves from Earth's surface to the sky and back again.

**Hygrometer** — Instrument used to measure humidity.

**Hydrosphere** — The water found on Earth.

**Iceberg** — Large chunk of ice that breaks off of a glacier in the melting process.

**Igneous rock** — A type of rock made under the lithosphere when old rocks melt, form magma, cool, and become solid again.

**Impermeable** — Cannot hold or be penetrated by water.

**Infiltration** — Process by which precipitation soaks through the soil and becomes groundwater.

**Inner core** — The very center of Earth; made up of solid iron.

**Last quarter moon** — In this phase, which follows the full moon, the Moon appears to be a half circle.

**Lava** — Magma that has been forced out of a volcano's vent to the outside of the volcano.

**Lava bombs** — Giant clouds of volcanic dust forced out of a volcano by a massive explosion.

**Lithosphere** — The part of Earth made up of the upper mantle and the crust.

**Lower mantle** — The area inside Earth that is solid rock and surrounds the core.

**Magma** — Combination of molten rock, gas, ash, and rock from inside Earth's crust.

**Magnitude** — Measure of strength or quality; an earthquake's magnitude is measured on the Richter scale.

**Maria** — Low spots or plains on the Moon that do not contain water but that are visible as large dark areas and were once thought to hold water. (*Maria* is the plural for *mare,* meaning "sea.")

**Mercury** — A heavy toxic liquid.

**Mesosphere** — Part of the temperature method of classifying the atmosphere, this layer is from 31 to 50 miles above Earth.

**Metamorphic rock** — A type of rock made when an existing rock is exposed to extreme temperature and pressure, causing the rock to change in form.

**Meteorites** — Meteors that did not burn up as they fell through our atmosphere and thus hit Earth.

**Meteorology** — The study of weather.

**Meteors** — Collections of dust that move through the sky so fast they burn.

**Milky Way Galaxy** — The galaxy in which Earth is located.

**Milne, John** — The scientist who invented the seismograph.

**New moon** — In this phase, the Moon appears totally black; we cannot see the sunlit portion.

**Nimbus** — Lower, dark clouds that contain water; name translated means "rain."

**Northern Hemisphere** — All of Earth above the equator.

**Oceanology** — The study of Earth's oceans.

**Orbit** — As a verb: To revolve around.

**Organic rock** — A type of sedimentary rock formed by plants or animals decomposing in water or mud.

**Outer core** — The area of Earth that surrounds the inner core; made up of liquid iron and nickel.

**Pangea** — A theory that says all of Earth's continents were once joined together and have moved apart over time and with the shifting of plates.

**Permeable** — Able to hold water or allow water to pass through.

**Photosphere** — Surface of the Sun.

**Piedmont glacier** — Glacier that spreads out at the foot of mountains, formed by the merging of two or more valley glaciers.

**Plate boundaries** — Places where one tectonic plate meets another.

**Plates** — In the theory of plate tectonics, large, thin, relatively rigid layers in Earth's crust that constantly move in relation to one another.

**Plate tectonics** — A theory that says Earth's crust is made up of several pieces called plates that are in constant motion and sometimes collide, forcing changes on Earth's surface.

**Precipitation** — Water that falls to Earth in the form of rain, snow, sleet, hail, or freezing rain.

**Psychrometer** — A type of hygrometer that measures the relative humidity.

**Recharging** — The process of precipitation infiltrating the ground and adding to the underground water supply.

**Relative humidity** — The amount of moisture the air can hold at a specific temperature.

**Revolve** — To move around another object.

**Richter scale** — Scale used to rate the strength of an earthquake.

**Rocks** — Natural mineral formations that make up most of Earth's crust.

**Rock cycle** — The process by which rocks are formed, altered, destroyed, and reformed by geological processes and which is recurrent, returning to a starting point.

**Rocky planets** — Planets composed mostly of rock and metal: Mercury, Venus, Earth, and Mars.

**Rotate** — To spin on an axis, either real or imaginary.

**Saffir-Simpson scale** — Scale used to rate the strength of a hurricane.

**Saltwater** — Water found in oceans, seas, and some lakes that contains high levels of salt and other minerals left behind by the evaporation process.

**San Andreas Fault** — 1,000-mile-long fault in Southern California; responsible for many of California's earthquakes.

**Satellite** — An object that orbits another object.

**Saturated** — Holding as much water as possible.

**Sea caves** — Caves formed when waves erode sea cliffs.

**Seasons** — Changes in weather patterns based on Earth's tilt toward the sun.

**Sediment** — Mineral or organic matter deposited by wind, water, or ice.

**Sedimentary rock** — A type of rock formed when loose sand, mud, and gravel deposited from moving water build up in layers and harden together.

**Seismogram** — A drawing, made by a seismograph, of Earth's vibrations.

**Seismograph** — Instrument that measures and records the shaking, or vibrations, of Earth's crust.

**Seismology** — The study of earthquakes.

**Shield volcano** — Nonexplosive volcanic mound with gentle, sloping sides, known to vary greatly in size.

**Sleet** — Begins as rain but becomes ice as it is falling to the ground. Occurs when the air temperature is above freezing but decreases closer to the ground.

**Snow** — A form of precipitation made up of many ice crystals grouped together that fall to the ground as snowflakes.

**Solar corona** — Outer atmosphere of the Sun; can be seen during an eclipse.

**Solar eclipse** — Occurs when the Moon's position blocks our view of the Sun.

**Solar system** — A group of planets and other heavenly bodies that revolve around the Sun.

**Soluble** — Able to be dissolved.

**Southern Hemisphere** — All of Earth below the equator.

**Speleothems** — Cave formations of various shapes and sizes made from mineral deposits.

**Spelunking** — The hobby of cave exploration.

**Stalactites** — Cave formations that hang from the ceilings of caves; made from mineral deposits.

**Stalagmites** — Cave formations that build up from the ground or floor of caves; made from mineral deposits.

**Star** — A heavenly body that is made of gas and produces its own light.

**Stratosphere** — Part of the temperature method of studying the atmosphere, this layer extends from about 7 miles to about 31 miles above Earth. It is just above the troposphere.

**Stratus** — Clouds formed by slowly rising air; can occur as high as 20,000 feet and last for days. Name translated means "layer."

**Sulfuric acid** — Acid formed by hydrogen sulfide mixing with oxygenated water; stronger than carbonic acid, hence it forms caverns much more quickly.

**Sun** — A star located closer to Earth than other stars. An enormous gaseous object in our solar system around which Earth and other planets revolve.

**Sunspots** — Areas where the surface of the Sun is cooler than the rest of its surface.

**Temperature** — One of two methods of studying the atmosphere; breaks the atmosphere into four layers, the troposphere, stratosphere, mesosphere, and thermosphere.

**Temperature** — How hot or cold the air is.

**Thermosphere** — Part of the temperature method of studying the atmosphere, this layer begins after the mesosphere.

**Thunder** — Sound created by the expansion of the air that is suddenly heated by lightning.

**Tornado** — Funnel cloud that comes down out of a storm cloud and actually makes contact with the ground.

**Torricelli, Evangelista** — Inventor of the barometer.

**Transpiration** — Process by which moisture in plants or trees turns to a vapor.

**Tropical cyclones** — Common term for hurricanes, cyclones, and typhoons.

**Tropical depression** — Storms organized into a group through strong winds and wind patterns.

**Tropical storm** — A tropical depression that has continued to grow and gain strength.

**Tropics** — The regions around the equator, characterized by consistently warm temperatures and warm ocean waters.

**Troposphere** — Part of the temperature method of studying the atmosphere, this is the lowest layer, extending almost seven miles above Earth.

**Typhoon** — Same as a hurricane and cyclone but occurring in a different part of the world.

**Upper mantle** — The area inside Earth that surrounds the lower mantle and is made of solid rock.

**Valley glacier** — Glacier often seen in river valleys, formed by the movement of alpine glaciers down mountain slopes.

**Vent** — Tube that connects the inside of a volcano to the deeper part of Earth's crust and the molten rock inside.

**Volcanic caves** — Caves formed by flowing lava and volcanic gases.

**Volcano** — Slope or mountain formed by magma.

**Volcanologist** — Scientist who studies volcanoes.

**Wall** — The edge of a hurricane's eye.

**Waning** — In this phase of the Moon, which follows the full moon, the Moon appears to get smaller.

**Waning crescent moon** — In this phase, the Moon appears as a sliver, just prior to the new moon.

**Waning gibbous moon** — In this phase, which follows the full moon, the Moon is on its way to becoming a half circle again.

**Warm front** — Warm air slowly rising as the cooler air sinks, causing condensation to take place.

**Water table** — Imaginary line underground, located just above the level of saturation.

**Waxing** — In this phase of the Moon, between the new moon and the full moon, the Moon appears to get larger.

**Waxing crescent moon** — In this phase, the sunlit portion of the Moon can be seen becoming larger each night, as the Moon and Earth change positions in relation to the Sun.

**Waxing gibbous moon** — In this phase, the Moon appears to be a little larger than a perfect half circle.

**Wet snow** — Snow that occurs when the air is just below freezing; good "packing" snow.

# Daily Reading Sheet

**Name** ______ **Unit/Lsn** ______ **Date** ______

Title, Author, Page #'s ______

______

______

**Main Topic** ______

______

______

I learned ______

______

______

I enjoyed learning about ______

______

______

I never knew that ______

______

______

**Facts** ______

______

______

______

I would like to know more about ______

______

**Vocabulary** ______

______

______

______

______

# Field Trip Journal

**Name** ________ **Unit/Lsn** ________ **Date** ________

**Destination** ________

**Purpose** ________

________

________

**I saw** ________

________

________

**I also saw** ________

________

________

**I learned** ________

________

________

The **most interesting** thing I saw (or learned) was ________

________

________

The most **unusual thing** I saw (or learned) was ________

________

________

I would like to have **learned more** about ________

________

________

I think I could learn about this by ________

________

________

________

# Checking It Out Experiment Form

**Name** ______________________ **Unit/Lsn** __________ **Date** __________

**Name of Experiment** ______________________

**Book(s) used** ______________________

______________________

______________________

**Objective** ______________________

______________________

______________________

**Hypothesis**

Based on what I have read and studied, I believe ______________________

______________________

______________________

**The Experiment**

To test out this theory, I plan to ______________________

______________________

______________________

Desired Result ______________________

______________________

Actual Result ______________________

______________________

❑ I am pleased with the result I received

❑ I am not pleased with the result I received

I believe this experiment would have gone better if ______________________

______________________

______________________

I learned ______________________

______________________

# Write About It!

## Worksheet for a Unit Wrap-Up Composition

1. Think. You just cannot begin a composition without doing the planning first.

2. Complete the form that follows. This will help you sort out the information from the unit. Avoid short, one-word answers. When I write to you, I try to imagine what you'll be thinking as you read. I don't try to take shortcuts, but I do think about keeping it simple. Check for proper spelling of science vocabulary now. It's easier to look it up right now.

3. After completing the form, take the information and put it in paragraph form. Make sure each sentence in a paragraph relates to the others and to the main idea of the paragraph.

4. Create an appropriate title for your composition. You can be creative, as I've been, or factual. ("Let's Go Spelunking" or "The Facts I Learned About Caves")

5. Be sure to add this nice wrap-up to your science notebook.

## Form

Unit No. ___________ Unit Title ______________________________________________

1. How many lessons are in the unit? ________________________________________

2. Make a list of the main topic of each lesson in the unit.

   ____________________________________________________________________

   ____________________________________________________________________

   ____________________________________________________________________

3. For each topic above, write at least three sentences that explain the information you covered. You can use the back of the form if this is not enough room. You might also include a sentence that describes the hands-on activity.

# Weather Observation Chart

| Outdoor Findings:<br><br>Weekday: | Temperature (Farenheit or Celsius) | Relative Humidity | Barometric Pressure | Sky Observation Sunny, Cludy? (Type) | Precipitation Present? What Form? | Rain Gauge Reading ___am/pm? |
|---|---|---|---|---|---|---|
| Monday | | | | | | |
| Tuesday | | | | | | |
| Wednesday | | | | | | |
| Thursday | | | | | | |
| Friday | | | | | | |
| Saturday | | | | | | |
| Sunday | | | | | | |

See instructions on the following page.

1. Remember, since your instruments are homemade, they are not totally accurate. You don't need to worry about that. The idea is to gather as much information as you can. This will help you understand the relationship of all these factors in the overall weather process.

2. You need to get your humidity readings from outside to be relevant to the weather outside.

3. You purchased a centigrade thermometer. It probably has a Fahrenheit reading as well. If you normally work in metric, I would use the Celsius reading; otherwise, use the Fahrenheit.

4. Official barometric readings are in terms that relate to sea level. Your reading, on the other hand, will be according to the inches on your ruler. This is okay. I just want you to know that your reading will not likely correspond to the one given in the official weather report.

5. Have Fun!

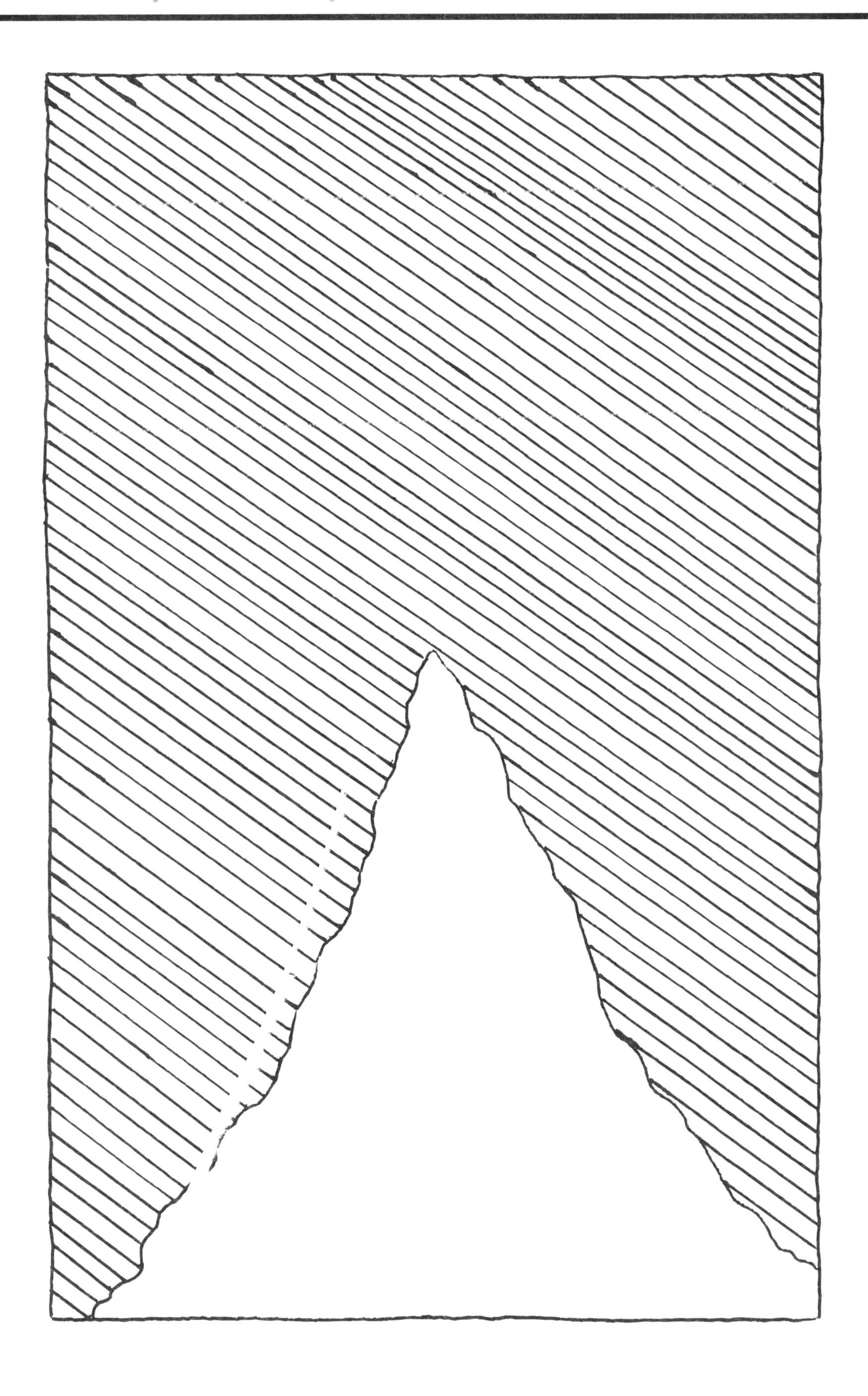

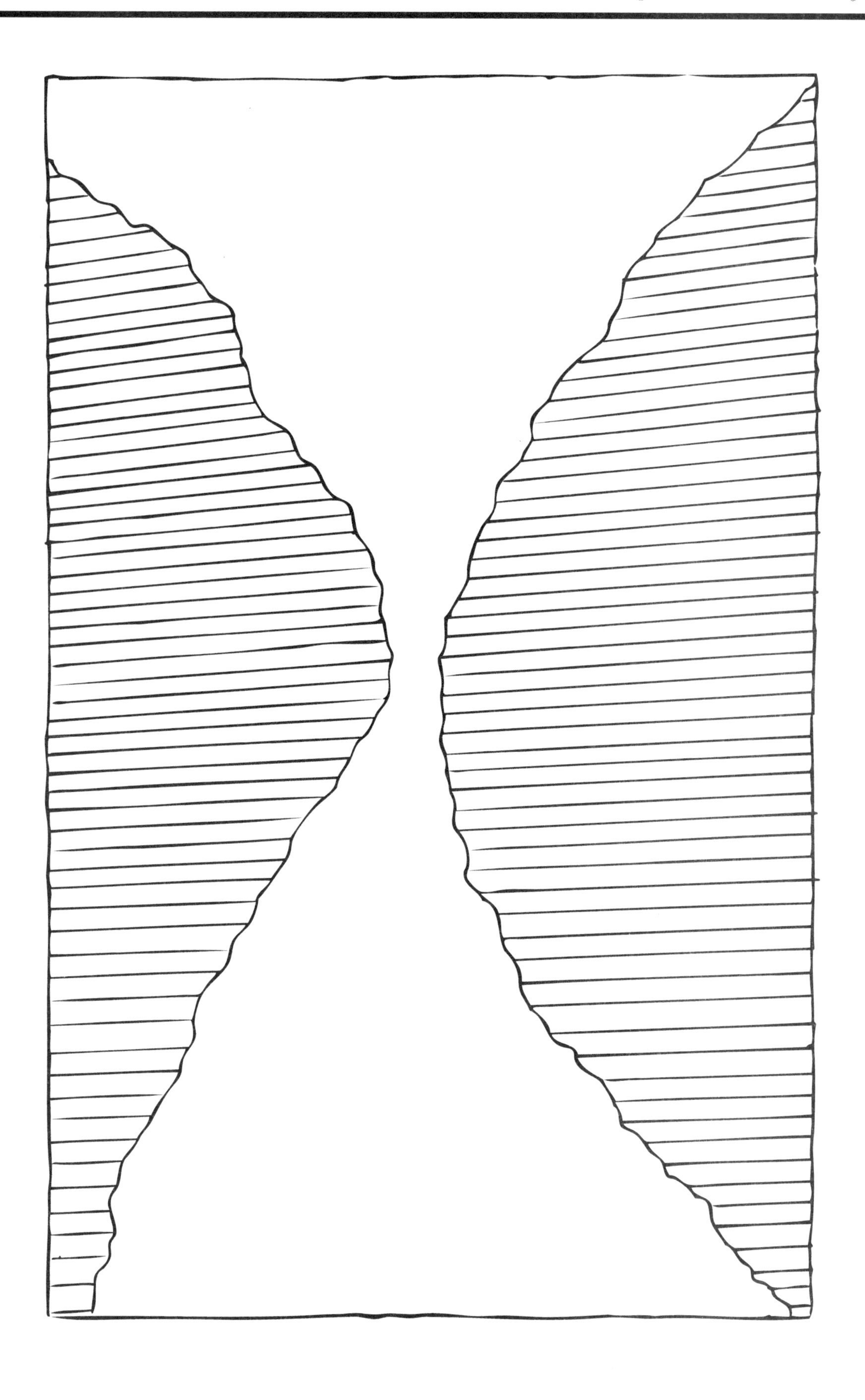

Mouth of a cave showing stalactite, stalagmite, and column formations

Mouth of a cave showing stalactite, stalagmite, and column formations

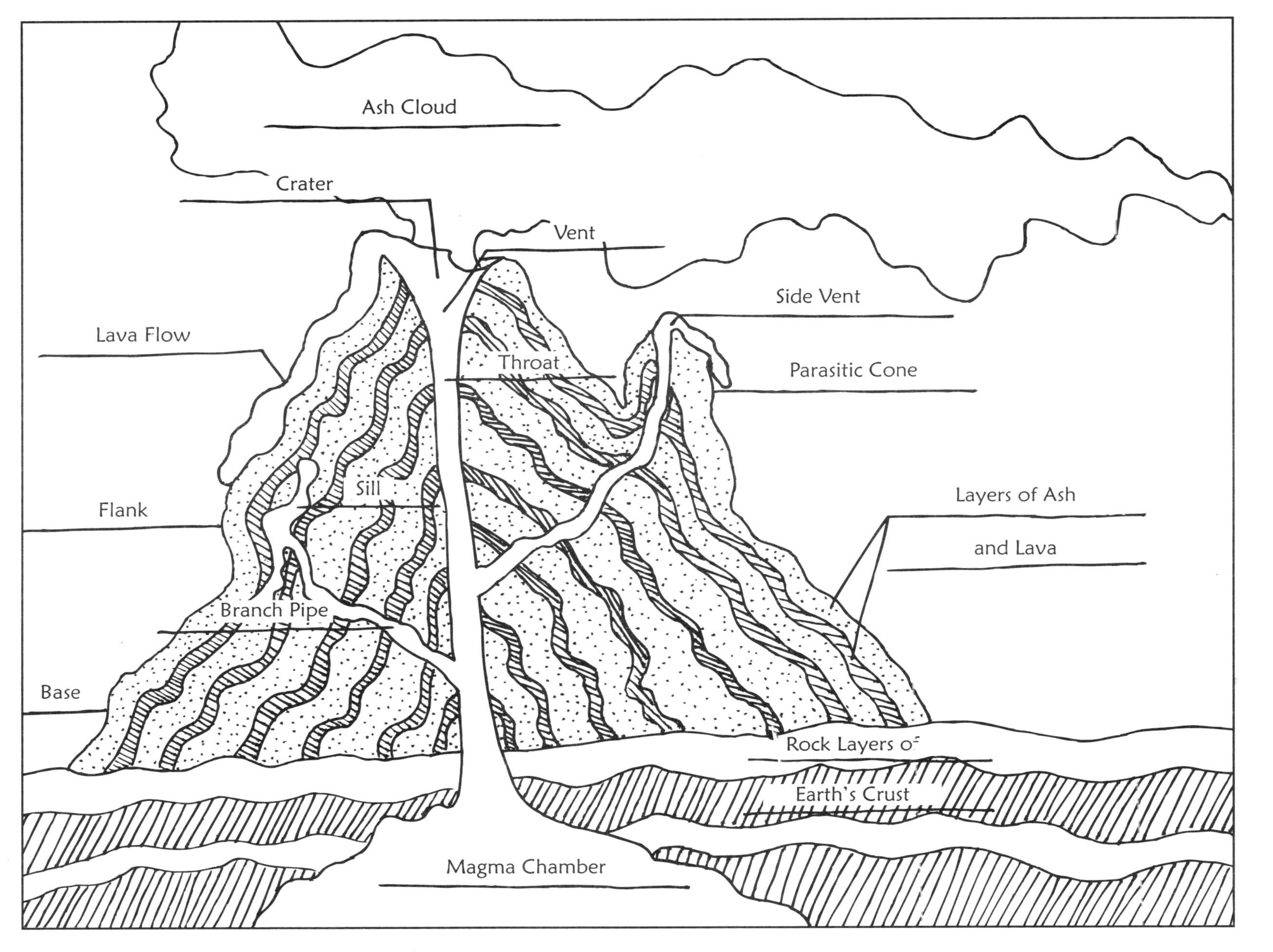


The Inside of a Volcano

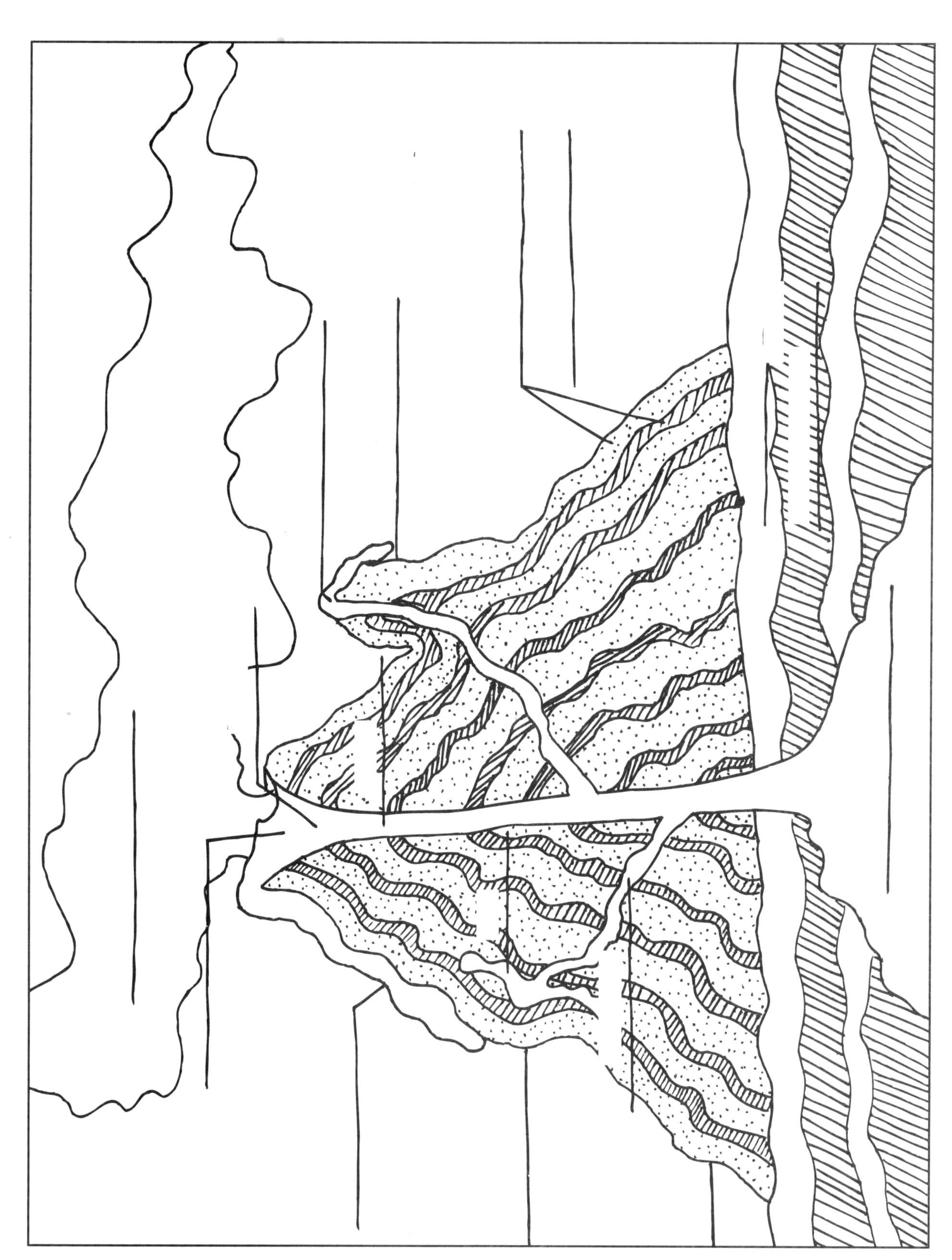

The Inside of a Volcano

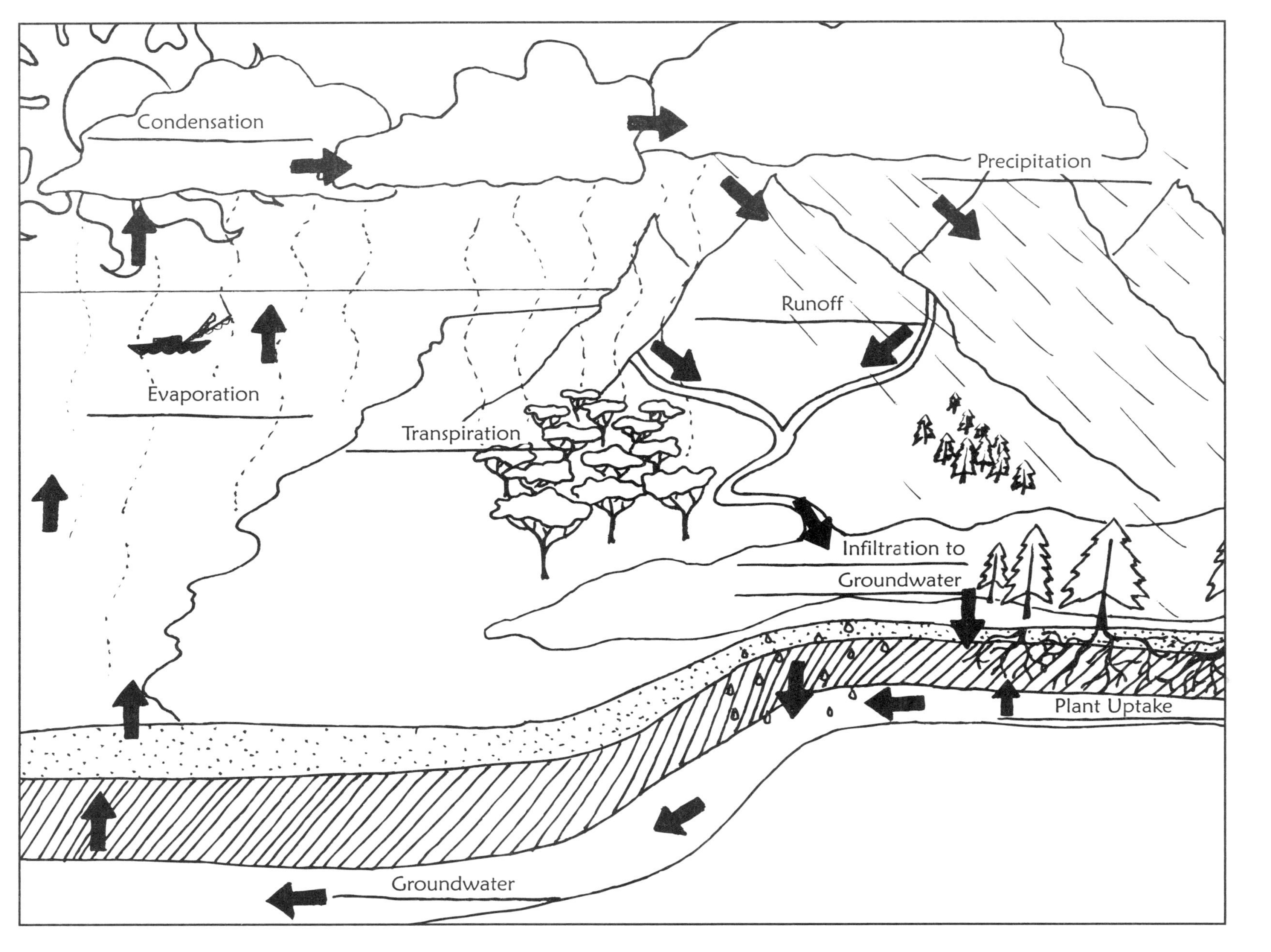


The Hydrologic Cycle

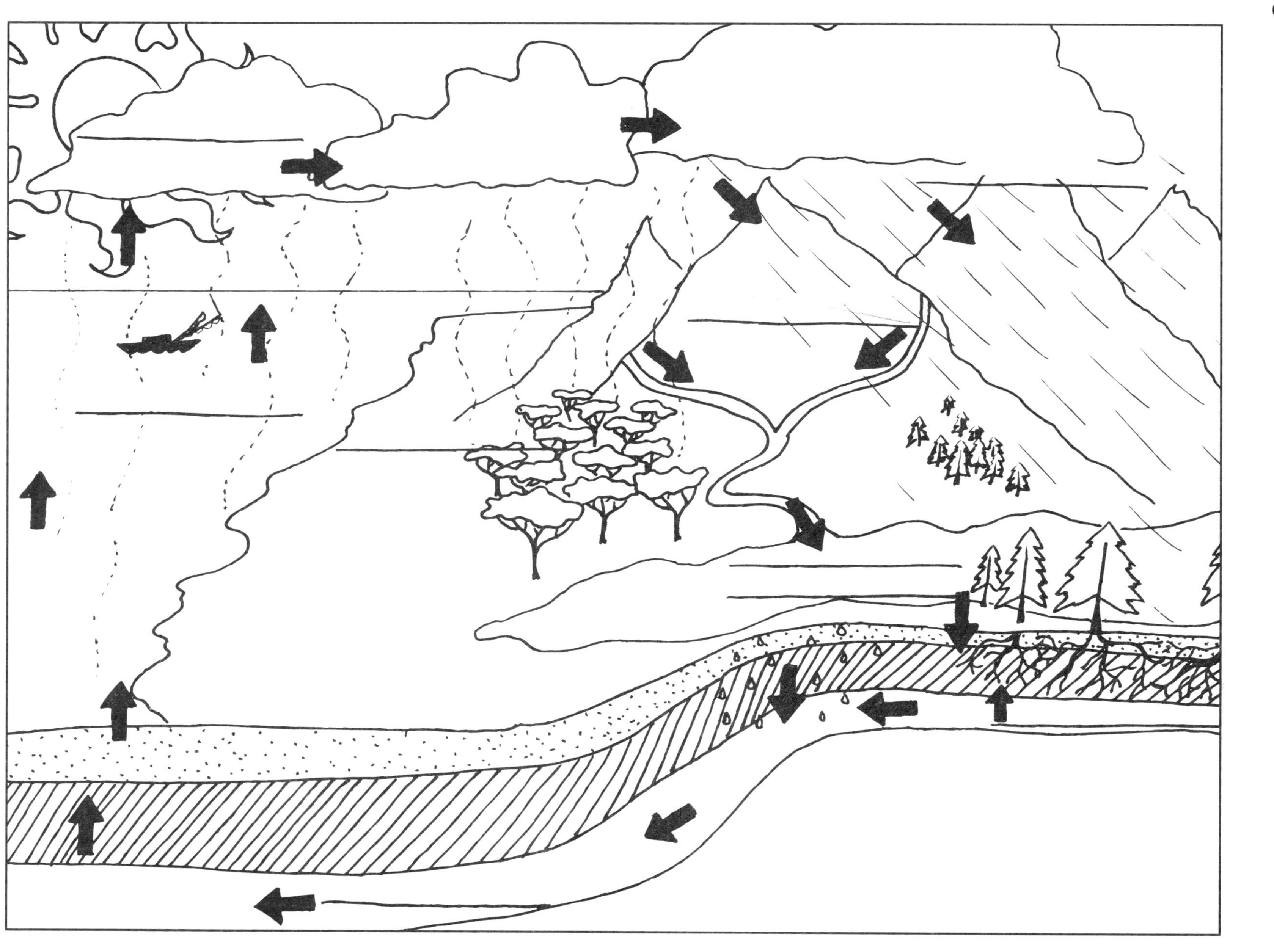

The Hydrologic Cycle

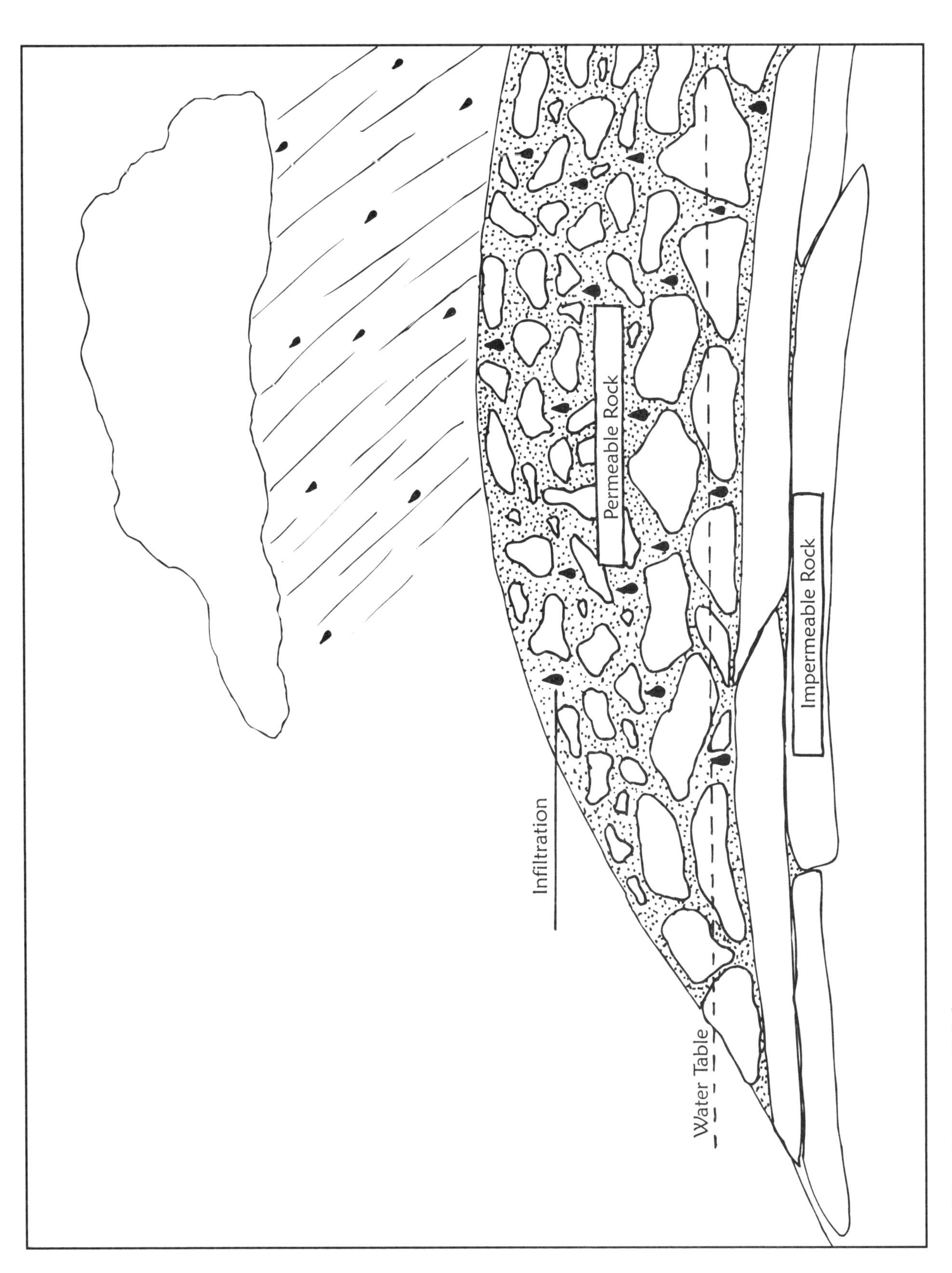


The Water Table

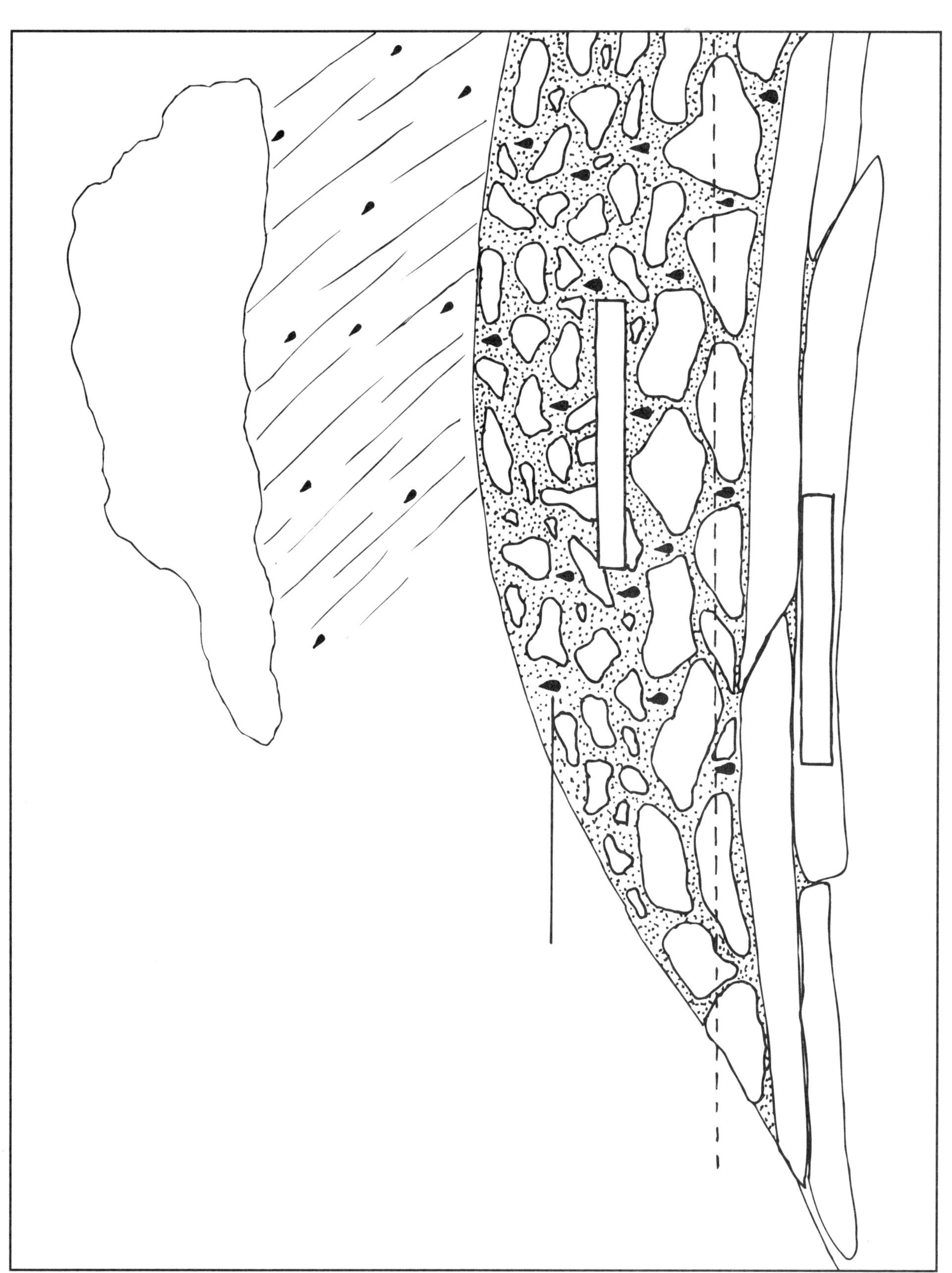

The Water Table

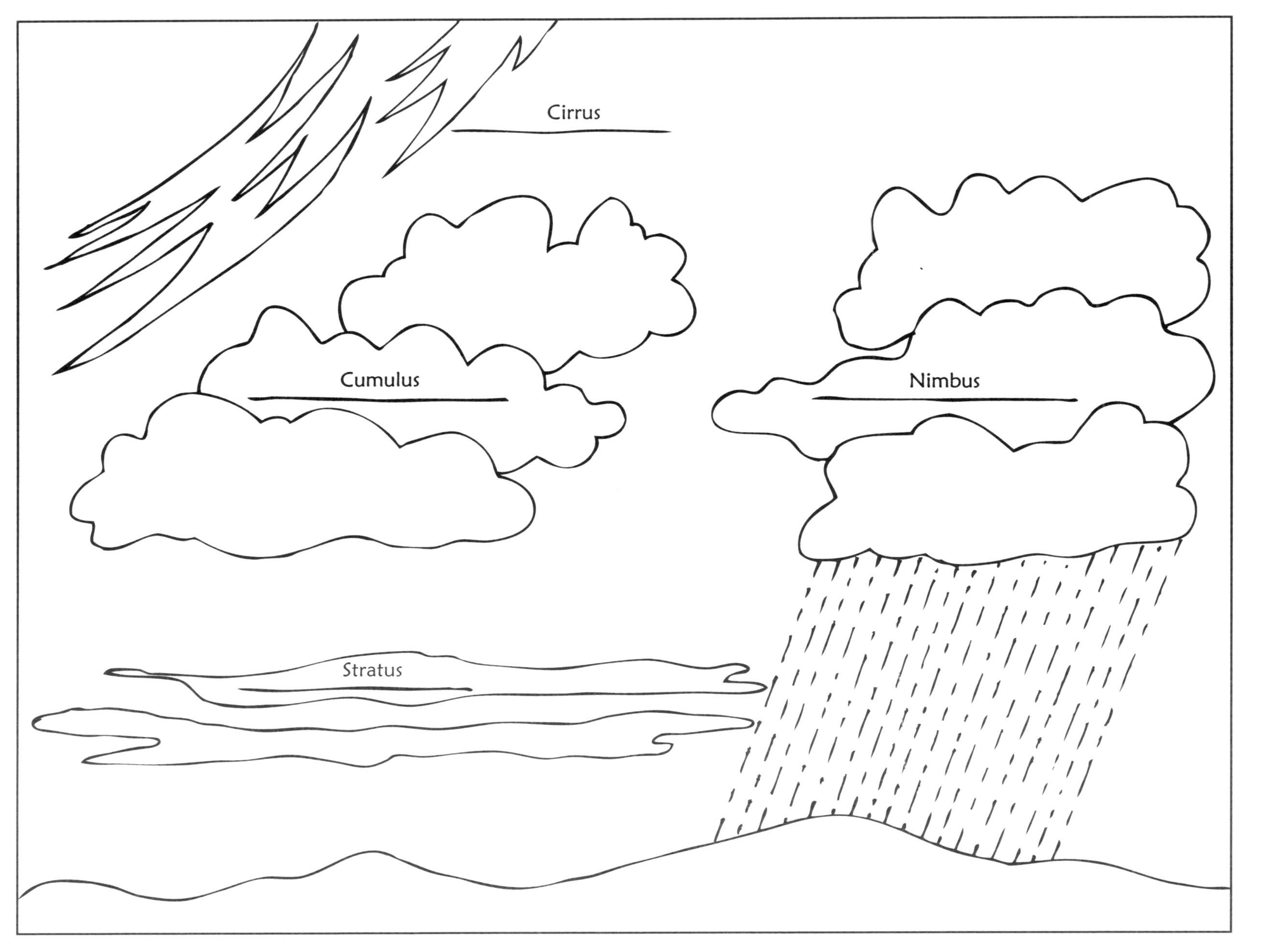


Types of Clouds

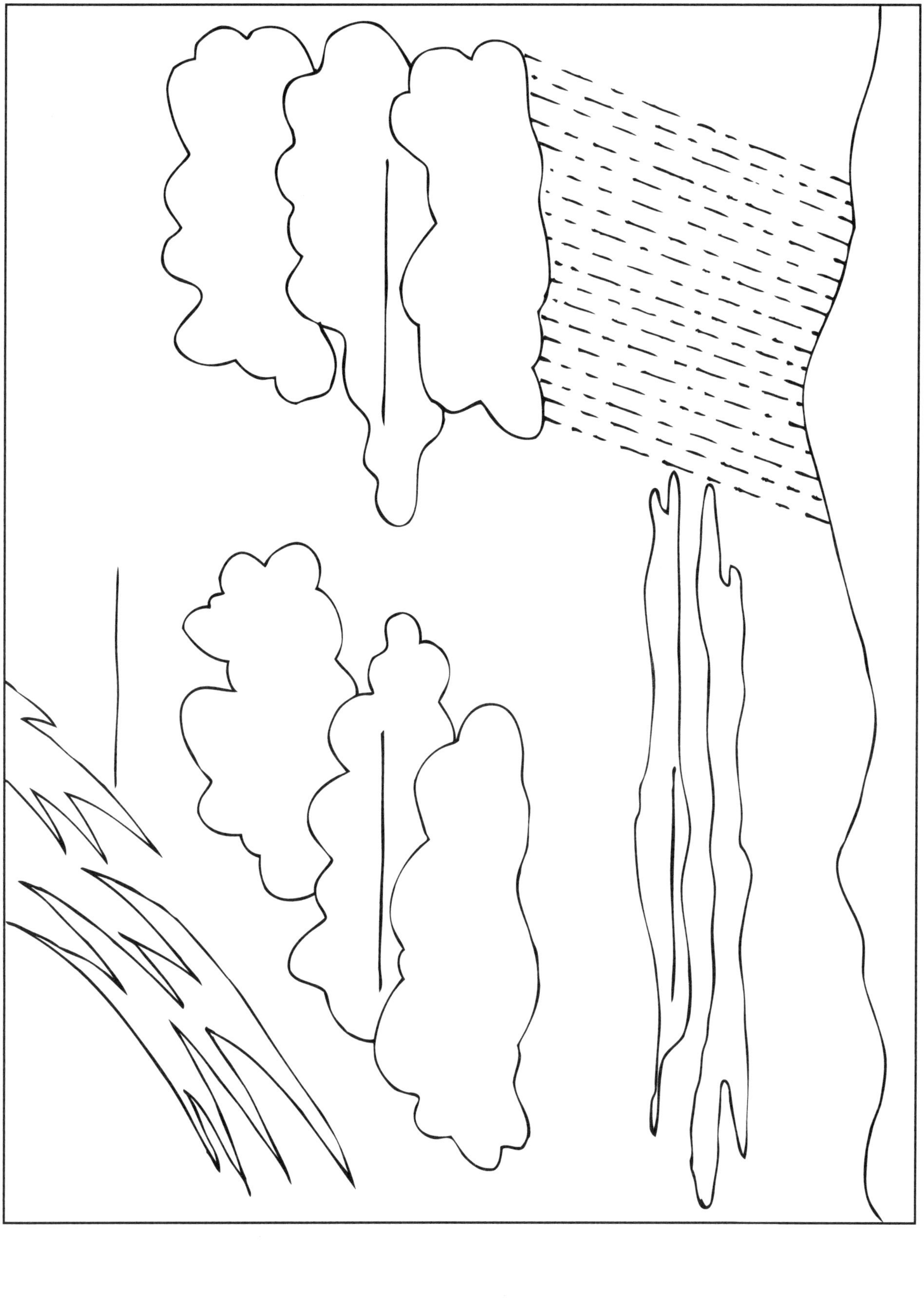

Types of Clouds

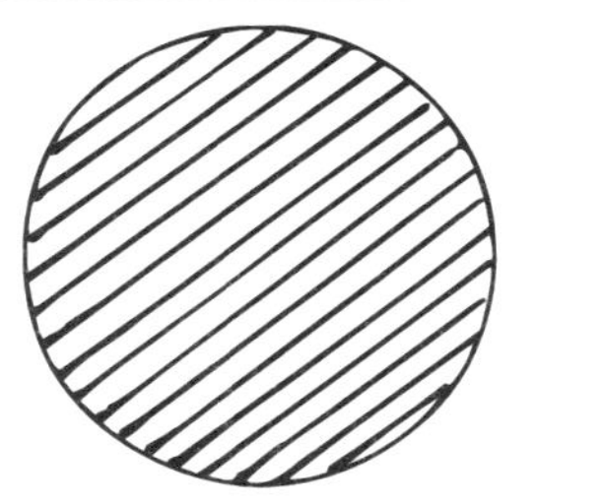

**New Moon**

The beginning of the cycle is called a **new moon**. A new moon looks totally black from Earth. We cannot see the sunlit portion at all.

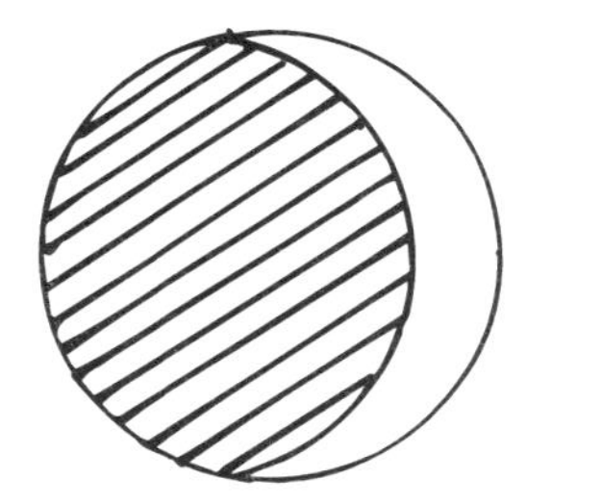

**Waxing Crescent Moon**

As the Moon and Earth change positions in relation to the Sun, we see a little more of the sunlit portion of the Moon each night. This is called a **waxing crescent moon**.

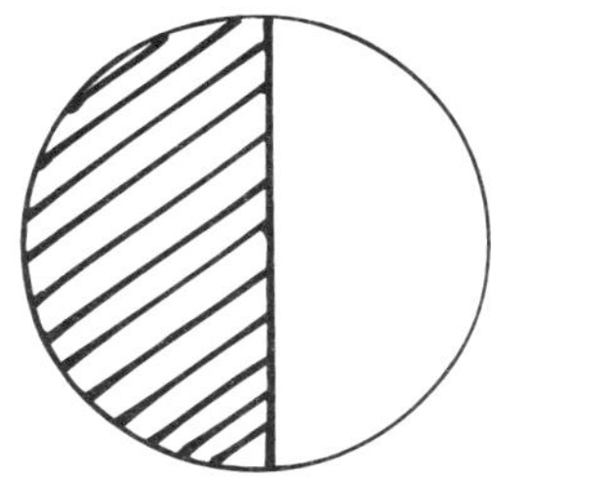

**First Quarter Moon**

Before long, we see a **first quarter moon**. A first quarter moon looks like a perfect half circle.

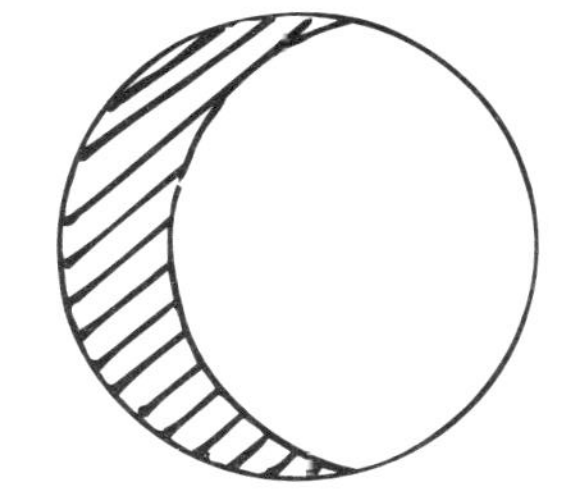

**Waxing Gibbous Moon**

When the Moon gets a little larger than a perfect half circle, it is known as a **waxing gibbous moon**.

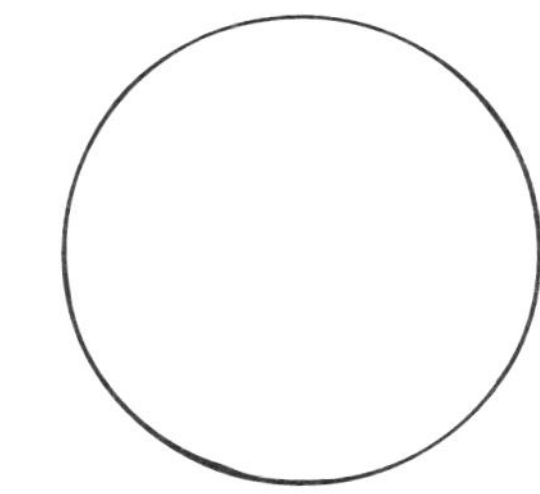

**Full Moon**

As the Moon continues its movement, the time comes when we can see the entire sunlit portion. This is called a **full moon**. This marks the halfway point.

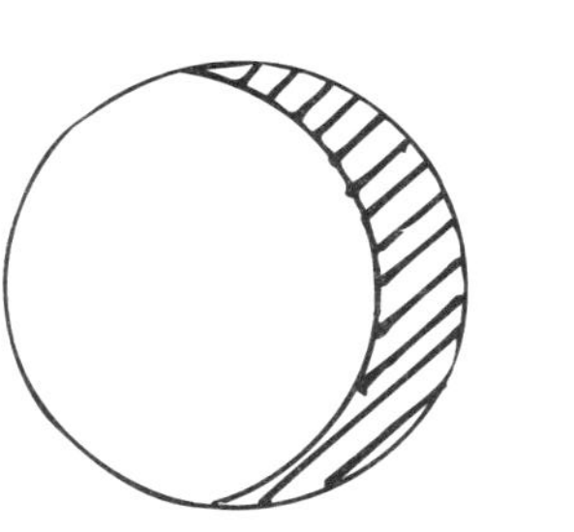

**Waning Gibbous Moon**

From this point on, Earth's view of the sunlit portion of the Moon gets smaller, so the Moon looks smaller. In this stage, the Moon is **waning**; on its way to a half circle again, it is a **waning gibbous moon**.

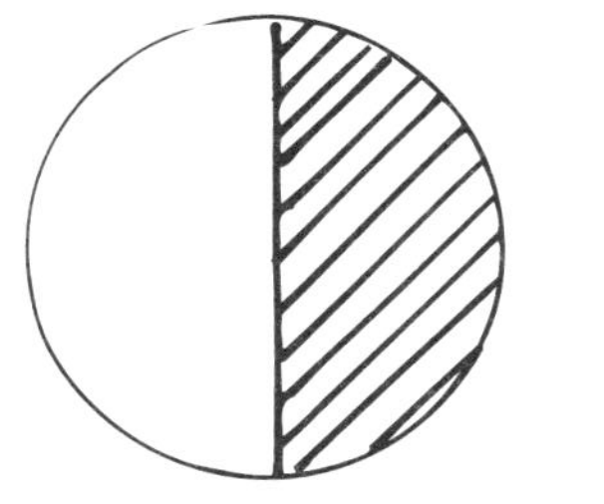

**Last Quarter Moon**

When we get to the half circle again, it is called a **last quarter moon**.

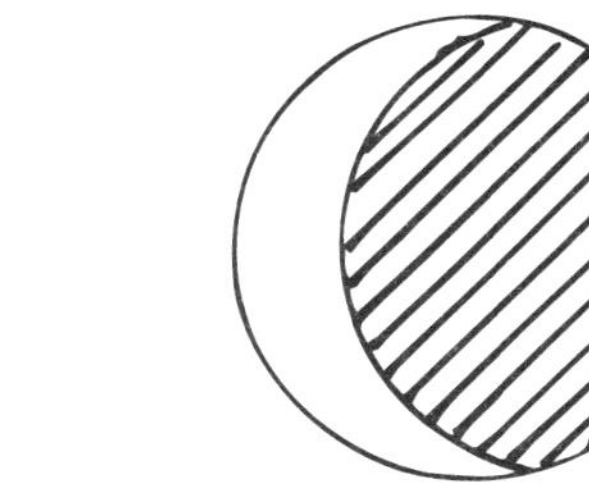

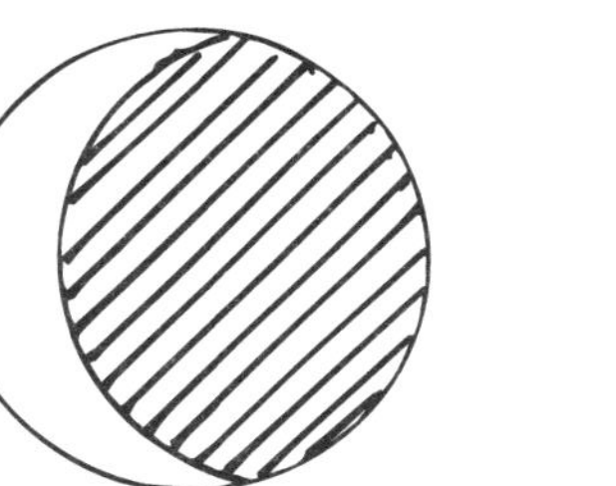

**Waning Crescent Moon**

The phase between a last quarter moon and a new moon is called a **waning crescent moon**. This looks like a sliver of a moon.

**New Moon**

Finally, the cycle is complete and we're back to a new moon.

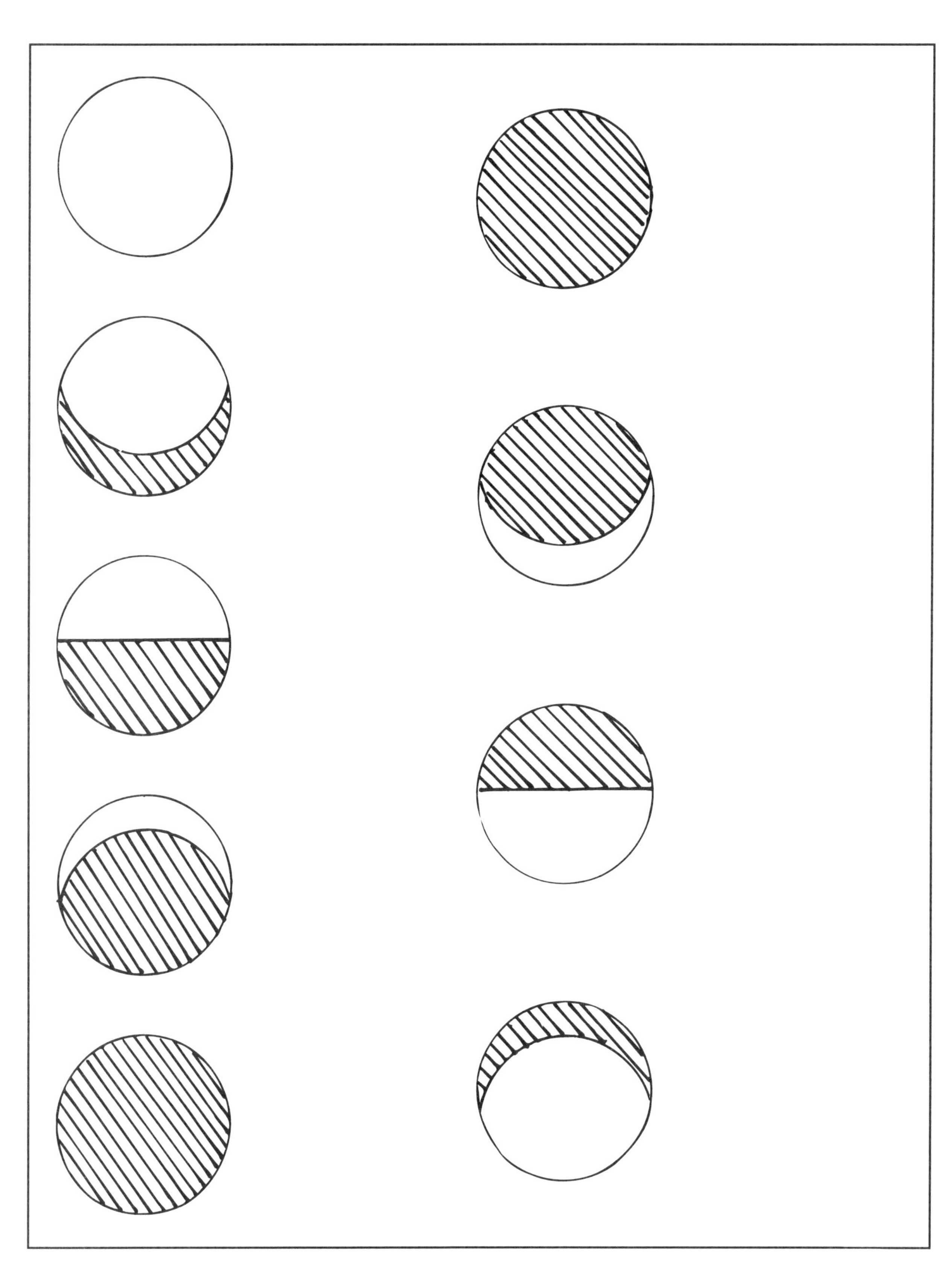

Stars
Solar System
moon

# UNIT THREE MEMORY LISTS

## Average Ocean Depths for Four Major Oceans

- Arctic Ocean — 3,407 feet
- Atlantic Ocean — 12,254 feet
- Indian Ocean — 12,740 feet
- Pacific Ocean — 13,740 feet

## Deepest Ocean Trenches

- Arctic Ocean — Eurasia Basin 17,881 feet
- Indian Ocean — Java Trench 25,344 feet
- Atlantic Ocean — Puerto Rico Trench 28,374 feet
- Pacific Ocean — Mariana Trench 36,200 feet

## Five Largest Lakes in the World

- Caspian Sea (Asia-Europe) — saltwater lake
- Superior (North America)
- Victoria (Africa)
- Aralb (Asia)
- Huron (North America)

## Five Longest Rivers in the World

- Nile (Africa)
- Amazon (South America)
- Yangtze (Asia)
- Mississippi-Missouri river system (North America)
- Yenisei-Angara river system (Asia)

# Unit Four Memory Lists

## Detailed Layers of the Atmosphere

| **Troposphere** | 0–9.5 miles | Weather occurs here. Jets fly here. |
|---|---|---|
| **Ozone Layer** | At about 12.5 miles | Protective layer. |
| **Stratosphere** | 9.5–31 miles | Contains the ozone layer. |
| **Mesosphere** | 31–50 miles | Shooting stars are seen within this layer. |
| **Thermosphere** | 50 miles and up | Aurorae occur at the base of this layer. |

# Unit Five Memory Lists

## Cloud Types from Lowest to Highest in the Sky

1. Stratus
2. Cumulus/nimbostratus/cumulonimbus (which build ever higher)
3. Stratocumulus
4. Altostratus/Altocumulus
5. Cirrocumulus
6. Cirrostratus
7. Cirrus

## Rainbow Formation

1. Sunlight is reflected by water droplets (often in clouds or rain).
2. The water bends or refracts the sunlight.
3. The different wavelengths of light are seen as individual colors.

## Colors of the Rainbow or Spectrum in Order

1. Red
2. Orange
3. Yellow
4. Green
5. Blue
6. Indigo
7. Violet

*Trick for remembering colors of the spectrum*: Think of the name ROY G. BIV. Each letter of the name represents a color of the spectrum. This name puts each color in its proper order.

# Unit Six Memory Lists

### Planets in Order from the Sun

1. Mercury
2. Venus
3. Earth
4. Mars
5. Jupiter
6. Saturn
7. Uranus
8. Neptune
9. Pluto

### Moon Cycle

1. New moon
2. Waxing crescent moon
3. First quarter moon
4. Waxing gibbous moon
5. Full moon
6. Waning gibbous moon
7. Last quarter moon
8. Waning crescent moon
9. New moon

### Planets from Largest to Smallest

1. Jupiter
2. Saturn
3. Uranus
4. Neptune
5. Earth
6. Venus
7. Mars
8. Mercury
9. Pluto

## Appendix C

# SCRIPTURE MEMORY

Scripture taken from New King James

Unit One
"In the beginning God created
the heavens and the earth."
(Genesis 1:1)

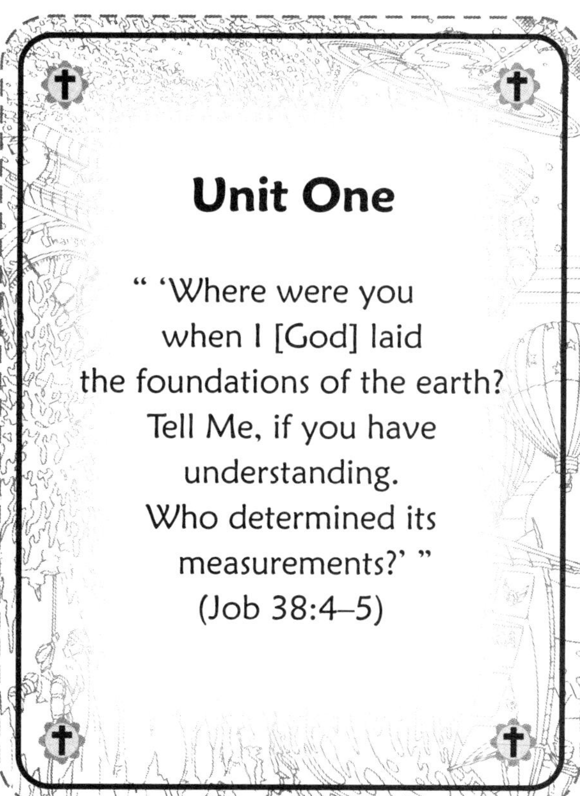
Unit One
" 'Where were you
when I [God] laid
the foundations of the earth?
Tell Me, if you have
understanding.
Who determined its
measurements?' "
(Job 38:4–5)

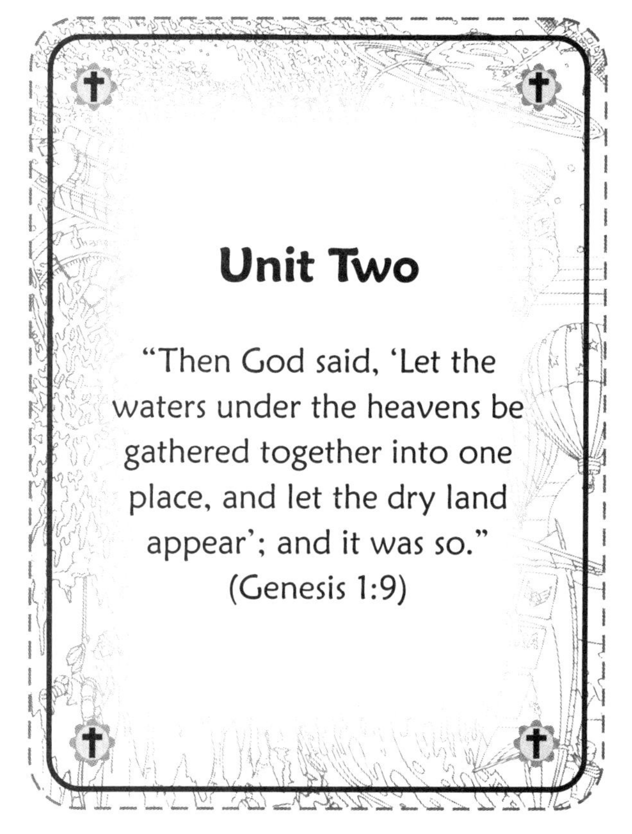
Unit Two
"Then God said, 'Let the
waters under the heavens be
gathered together into one
place, and let the dry land
appear'; and it was so."
(Genesis 1:9)

Unit Two
"So he [Saul] came to the
sheepfolds by the road, where
there was a cave; and Saul
went in to attend to his needs.
(David and his men were
staying in the recesses
of the cave.)"
(1 Samuel 24:3)

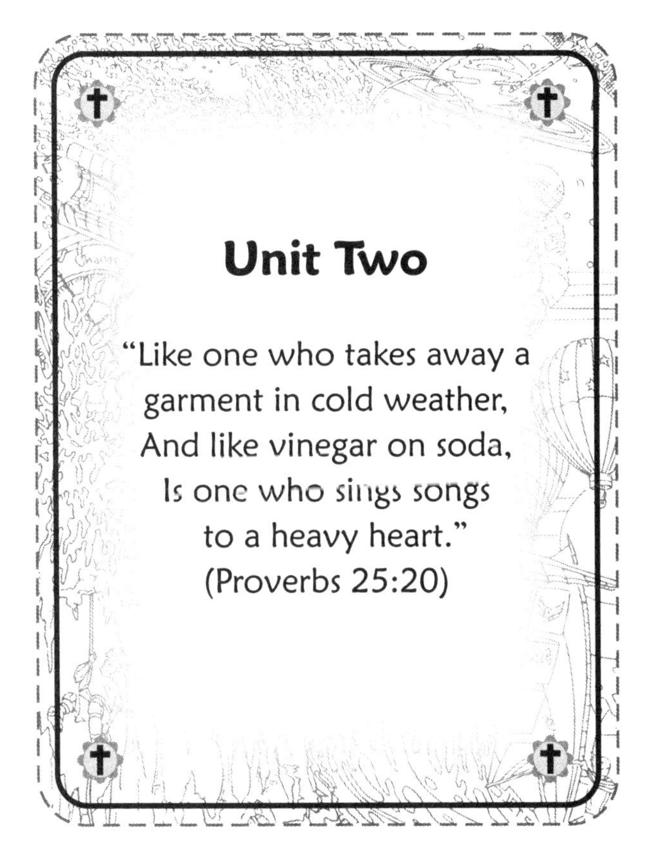

## Unit Two

"Like one who takes away a
garment in cold weather,
And like vinegar on soda,
Is one who sings songs
to a heavy heart."
(Proverbs 25:20)

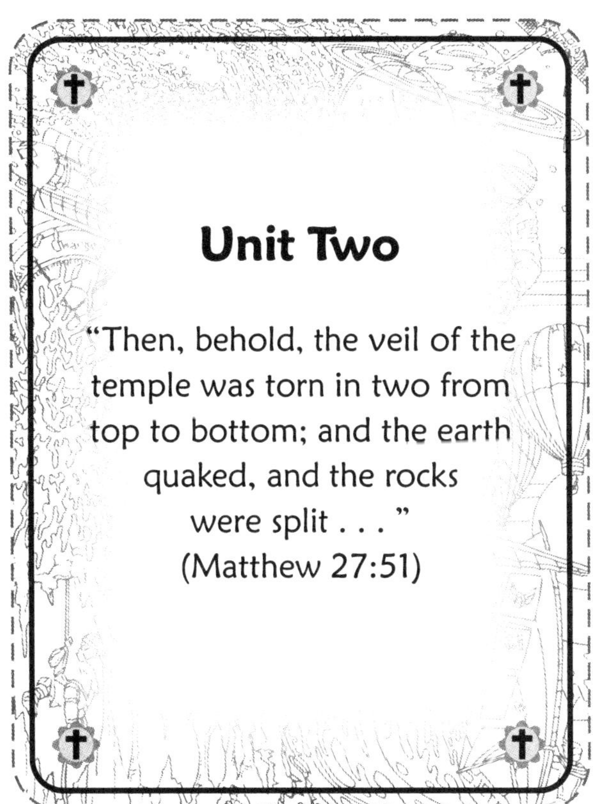

## Unit Two

"Then, behold, the veil of the
temple was torn in two from
top to bottom; and the earth
quaked, and the rocks
were split . . . "
(Matthew 27:51)

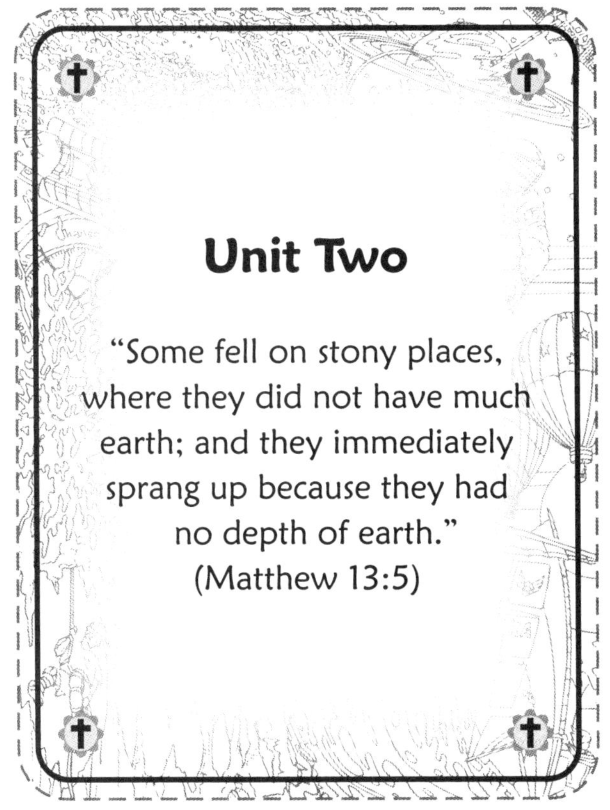

## Unit Two

"Some fell on stony places,
where they did not have much
earth; and they immediately
sprang up because they had
no depth of earth."
(Matthew 13:5)

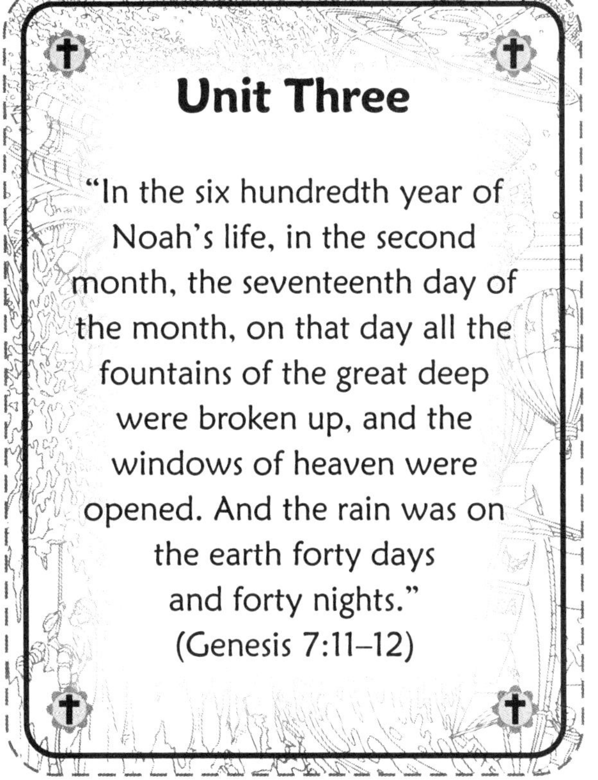

## Unit Three

"In the six hundredth year of
Noah's life, in the second
month, the seventeenth day of
the month, on that day all the
fountains of the great deep
were broken up, and the
windows of heaven were
opened. And the rain was on
the earth forty days
and forty nights."
(Genesis 7:11–12)

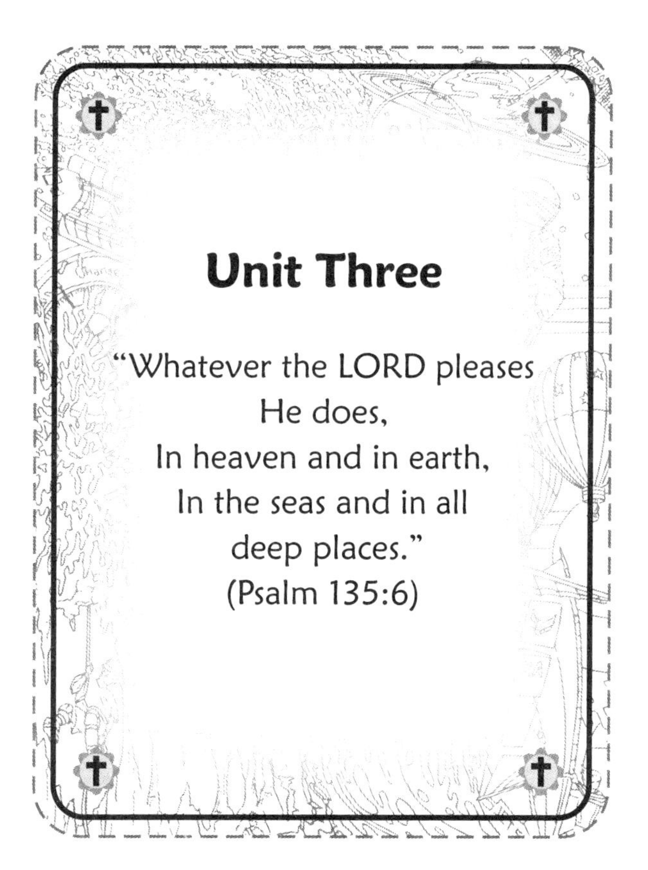

## Unit Three

"Whatever the LORD pleases
He does,
In heaven and in earth,
In the seas and in all
deep places."
(Psalm 135:6)

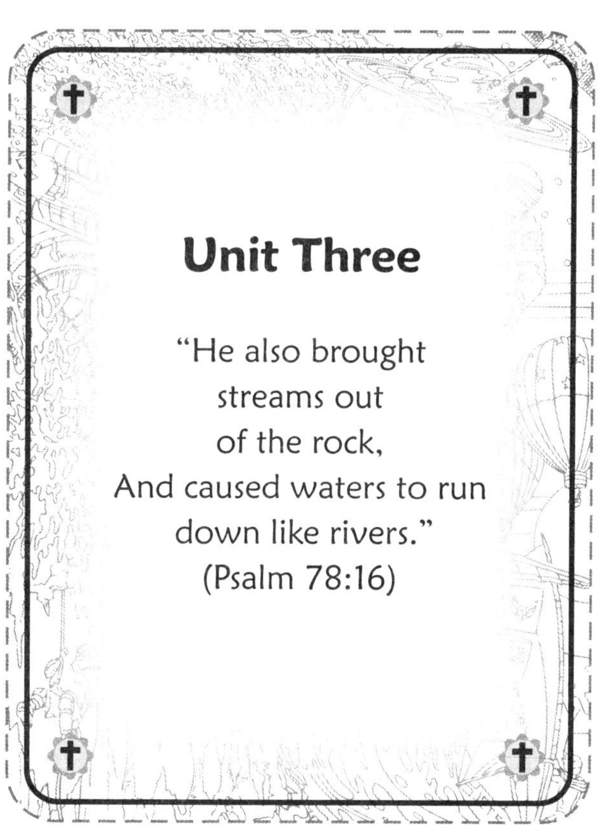

## Unit Three

"He also brought
streams out
of the rock,
And caused waters to run
down like rivers."
(Psalm 78:16)

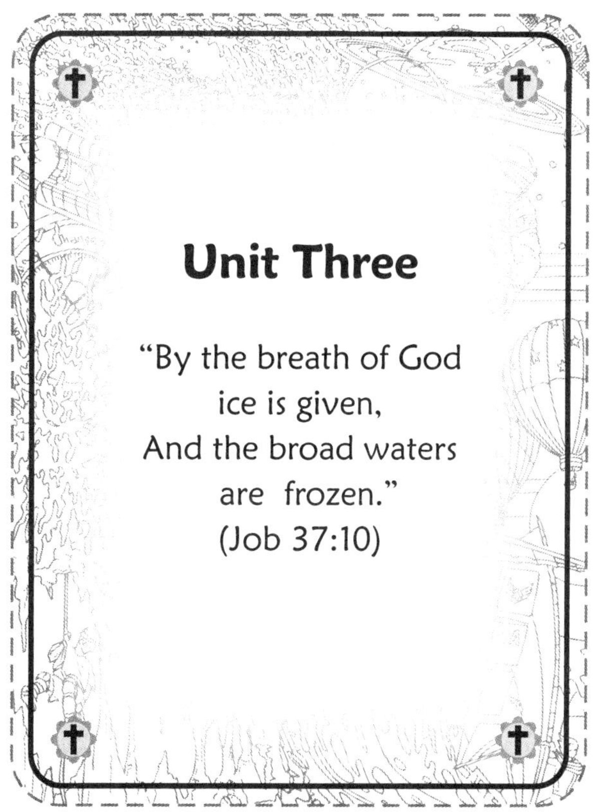

## Unit Three

"By the breath of God
ice is given,
And the broad waters
are frozen."
(Job 37:10)

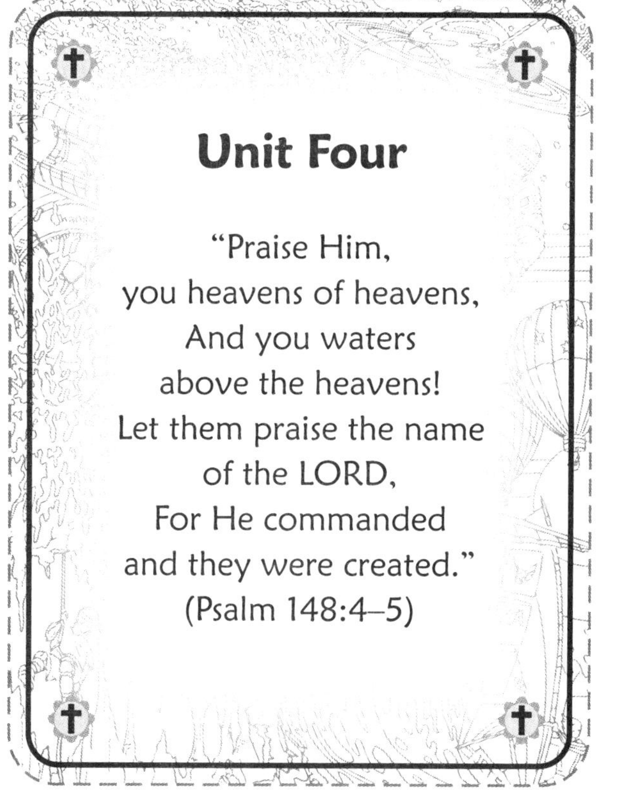

## Unit Four

"Praise Him,
you heavens of heavens,
And you waters
above the heavens!
Let them praise the name
of the LORD,
For He commanded
and they were created."
(Psalm 148:4–5)

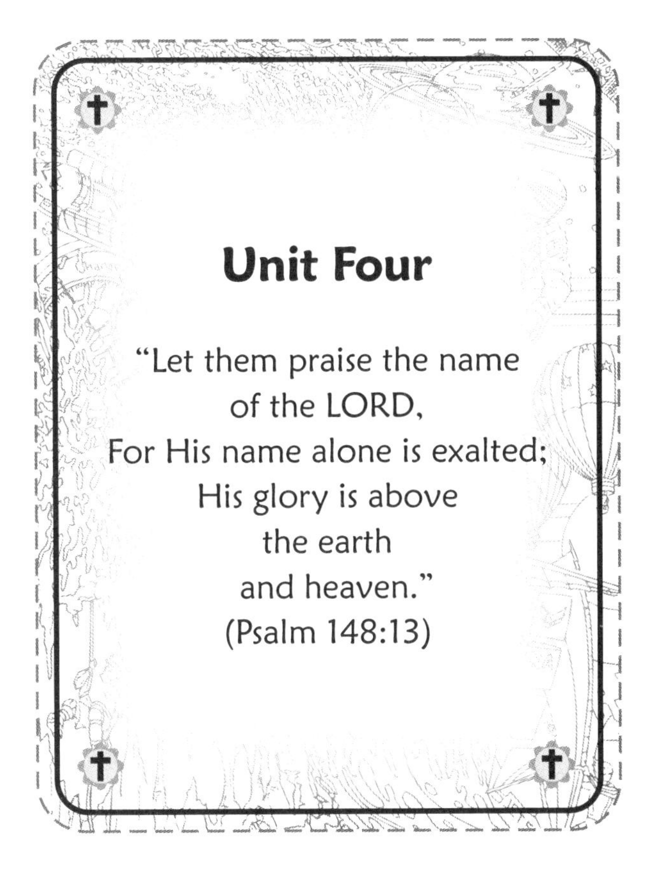
Unit Four
"Let them praise the name
of the LORD,
For His name alone is exalted;
His glory is above
the earth
and heaven."
(Psalm 148:13)

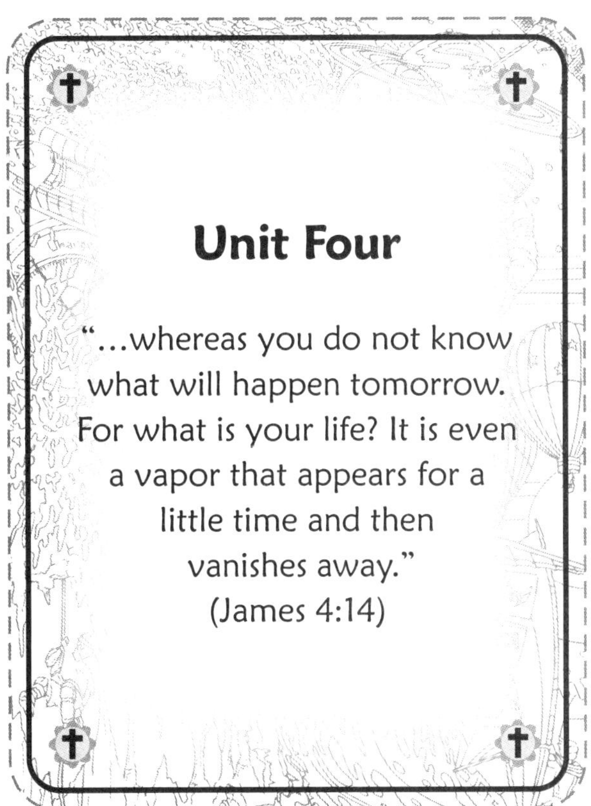
Unit Four
"...whereas you do not know
what will happen tomorrow.
For what is your life? It is even
a vapor that appears for a
little time and then
vanishes away."
(James 4:14)

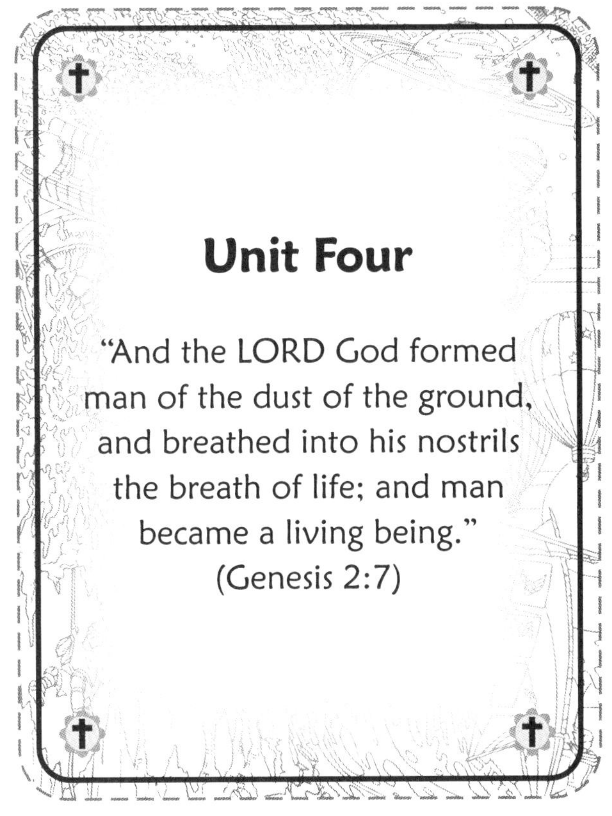
Unit Four
"And the LORD God formed
man of the dust of the ground,
and breathed into his nostrils
the breath of life; and man
became a living being."
(Genesis 2:7)

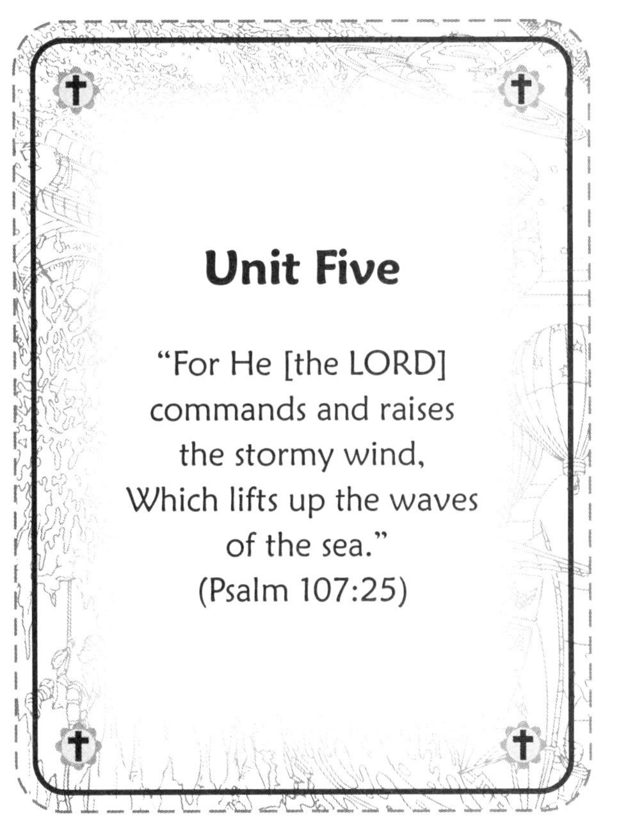
Unit Five
"For He [the LORD]
commands and raises
the stormy wind,
Which lifts up the waves
of the sea."
(Psalm 107:25)

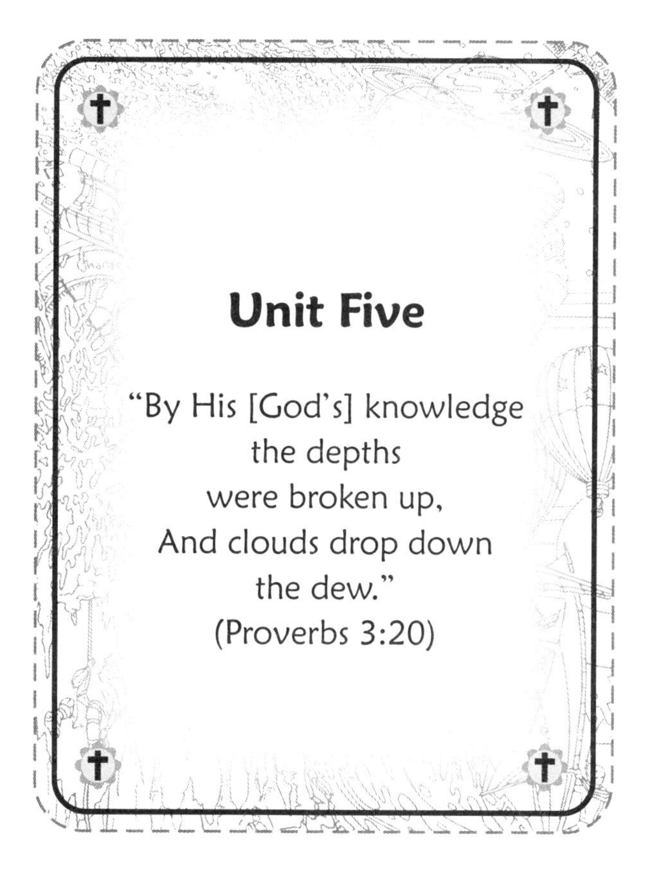
Unit Five
"By His [God's] knowledge
the depths
were broken up,
And clouds drop down
the dew."
(Proverbs 3:20)

Unit Five
"Praise the LORD
from the earth,
You great sea creatures
and all the depths;
Fire and hail,
snow and clouds;
Stormy wind, fulfilling His
word . . . "
(Psalm 148:7–8)

Unit Five
"He causes the vapors to
ascend from the ends
of the earth;
He makes lightning
for the rain;
He brings the wind out
of His treasuries."
(Psalm 135:7)

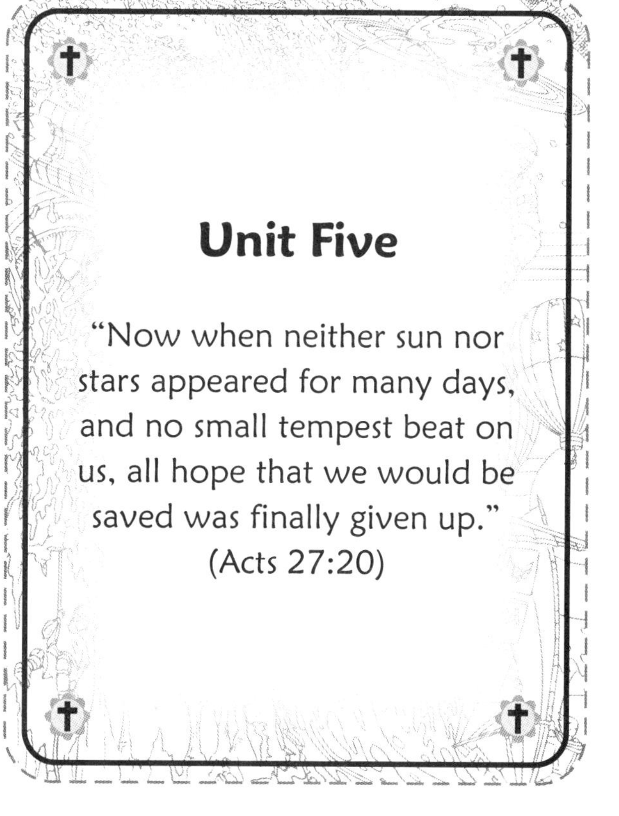
Unit Five
"Now when neither sun nor
stars appeared for many days,
and no small tempest beat on
us, all hope that we would be
saved was finally given up."
(Acts 27:20)

Unit Six
"The LORD by wisdom
founded the earth;
By understanding He
established the heavens . . . "
(Proverbs 3:19)

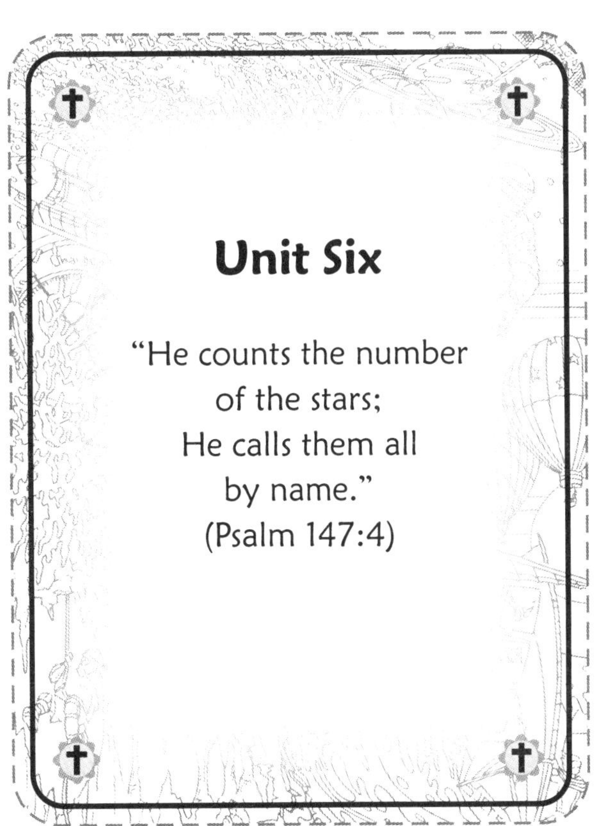
Unit Six
"He counts the number
of the stars;
He calls them all
by name."
(Psalm 147:4)

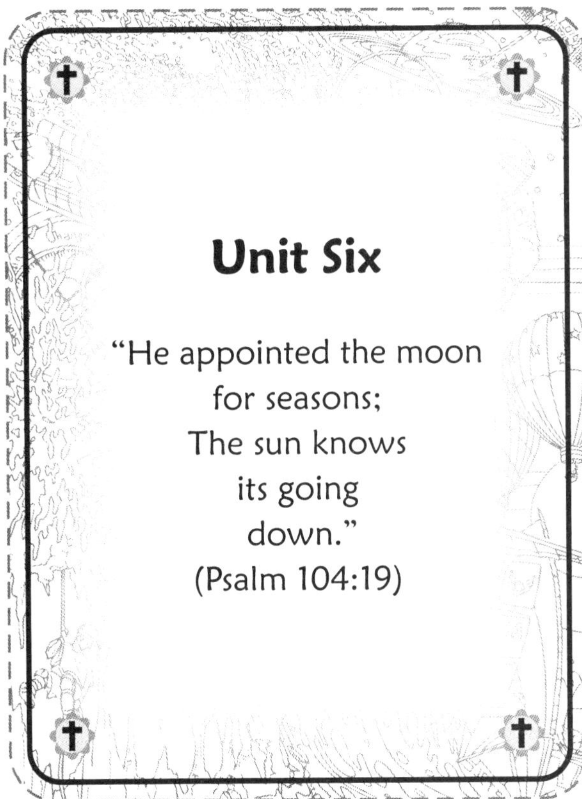
Unit Six
"He appointed the moon
for seasons;
The sun knows
its going
down."
(Psalm 104:19)

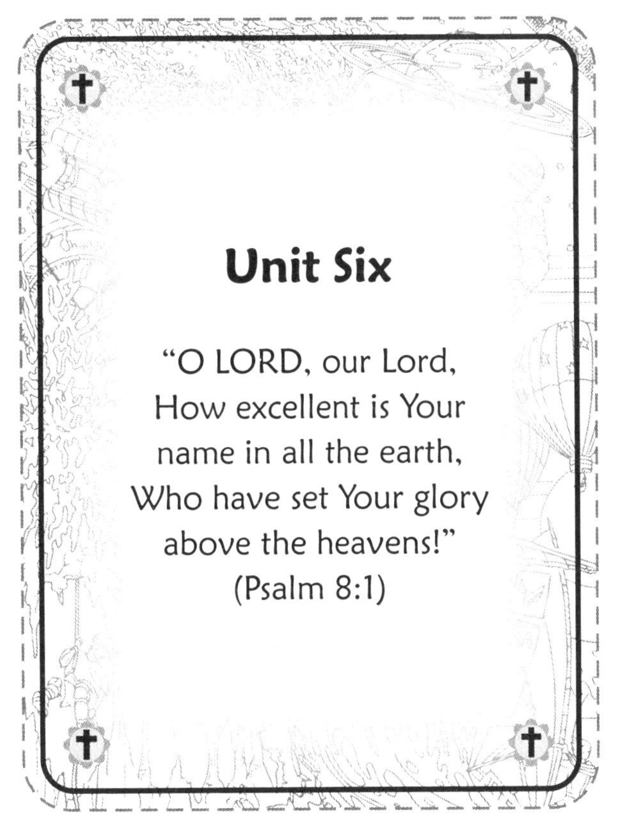
Unit Six
"O LORD, our Lord,
How excellent is Your
name in all the earth,
Who have set Your glory
above the heavens!"
(Psalm 8:1)

Appendix D

# COLORING PAGES

Coloring pages begin on the following page.

Change
Fair
Very Dry
KA
JET SET

OCEAN RESEARCH II
RIVER EXPLORATION
JETSET

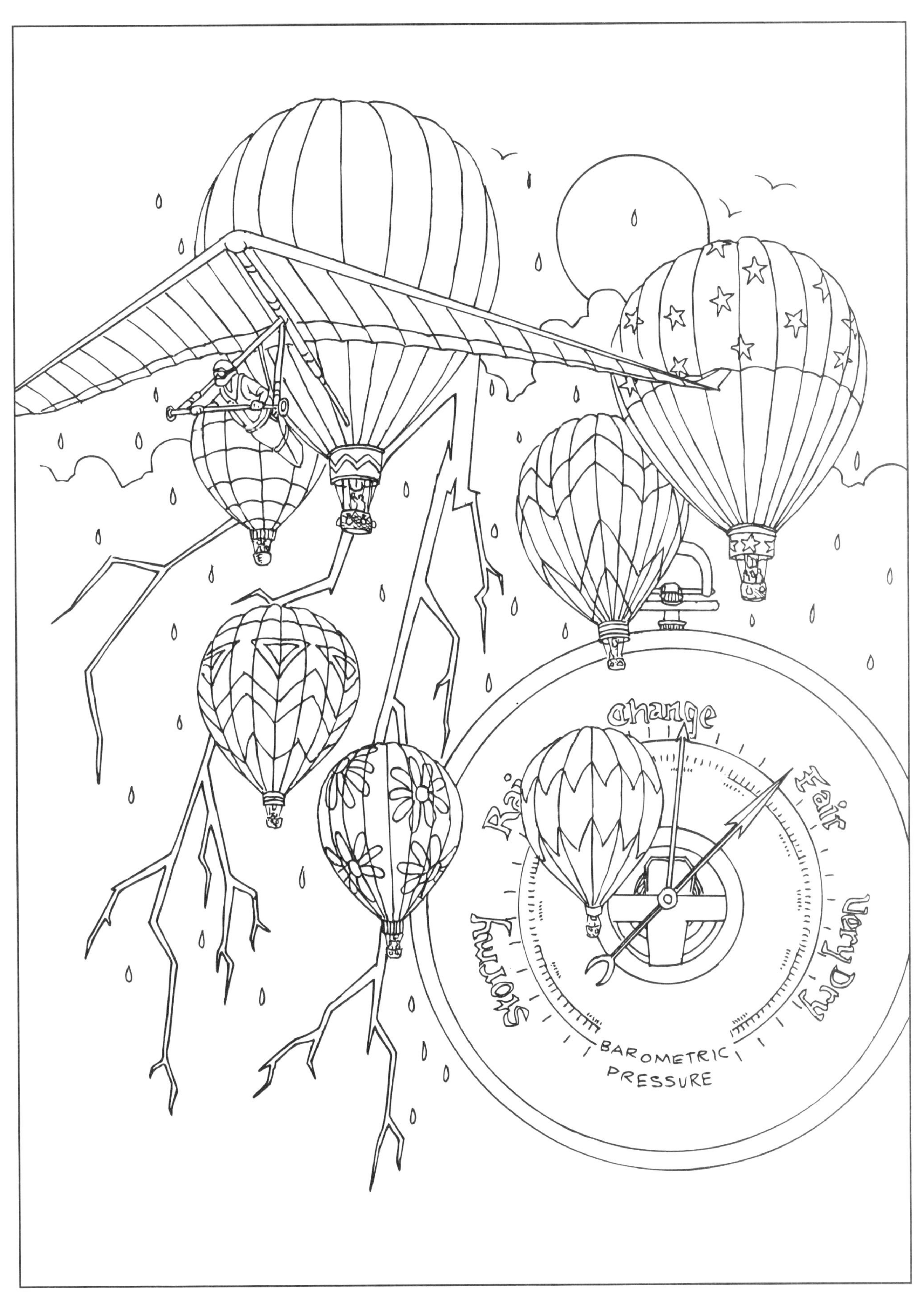
change
Rain
Fair
Stormy
Very Dry
BAROMETRIC
PRESSURE

NASA
NASA

# RECIPES AND SUPPLEMENTAL ACTIVITIES

## Supplemental Activity 1:

### Hands-On Time: Make Chocolate Nut Layer Bars

**Objective:**

To visualize the layers of sediment that combine to make one rock.
*Adult supervision required.*

**Materials**

- 1 stick butter, softened
- ¼ cup powdered sugar
- 1 teaspoon vanilla
- 1 cup all-purpose flour
- 1¼ cups sugar, granulated
- 1 cup whipping cream
- 2 sticks butter, sliced into 8 slices each
- 1 cup toasted nuts of your choice (Mixed is recommended: pecans and almonds.) For this exercise, chop nuts into varying sizes to simulate rocks.
- 6-ounce package semisweet chocolate chips

## Instructions

### Make the Crust

1. Beat stick of softened butter until creamy. Slowly add powdered sugar, beating well. Add vanilla and beat well. Gradually add flour, beating constantly at low speed until well incorporated.
2. Coat fingertips with flour and press mixture into an 8-X-8-inch, square baking pan. Bake at 325 degrees for 25 minutes. Crust should be golden brown.

### Make the Caramel

3. In a medium skillet, cook the granulated sugar over medium heat until it is caramel colored. Lower the heat and slowly add ⅔ cup of whipping cream and stir to combine. *Parents must do this part to protect children from spattering of hot sugar.*
4. Add the remaining butter, 4 pieces at a time. Stir well after each addition. Mixture should be smooth.
5. Remove from heat and let cool for 10 minutes.

### Put It Together

6. Sprinkle the nuts over the crust.
7. Pour the caramel over the crust.
8. Cool thoroughly.

### Add the Chocolate

9. Place remaining whipping cream in a microwave-safe container. Microwave on High for 1 minute.
10. Add the chocolate chips, and let sit for 2 minutes.
11. Whisk the chocolate and cream together until smooth.
12. Pour over the cooled caramel mixture. Spread the chocolate to edges of pan.
13. Cover and place in refrigerator. When layers are firm, the bars can be cut. Let stand at room temperature for 5 minutes before cutting.

# Supplemental Activity 2

## Hands-On Time: Make Rock Snacks

### Objective:

To visualize the layers of sediment that combine to make one rock. Try this alternative recipe for visualizing (and enjoying!) sedimentary rock layers. Have fun!
*Adult supervision required.*

### Materials

- ½ cup butter
- 1½ cups vanilla wafer crumbs
- 14-ounce can sweetened condensed milk
- 6-ounce package semisweet chocolate chips
- 6-ounce package peanut butter chips
- 1 cup chopped walnuts or pecans

### Instructions

1. Melt the butter in a small pan and pour into a clear 13-X-9-inch baking pan.
2. Sprinkle crumbs over the butter.
3. Pour condensed milk evenly over the crumbs.
4. Layer the remaining ingredients evenly over the top.
5. Press down gently. Bake at 350 degrees for 25–30 minutes.
6. Let cool. Cut into bars. Enjoy!

---

Adapted from http://library.thinkquest.org/J002289/snacks.html

# Supplemental Activity 3

## Atmosphere Vocabulary

H M N S M M Z J W X D D T L E N D X

E E P T R O P O S P H E R E Z Y F O

T X B P O K F O A D N M R S F T E B

E D P G L T A Z D E C E E Q L O X K

R N B A D M E B G Z H J A Q W R A O

O G U A E S L Y O P H E F R B R N Y

S A V R R X X W S U O D H I C I N L

P K H E S O R O U E B O Y X E C A I

H A E T I R M T E K L E N D R E O S

E N B E P O S E E W K I R I D L P Y

R E A M H X N G T R E B E L A L T S

E C R O P O Q K I E A S U B T I E B

N I T R O G E N H I R Z F A D C I Y

A H P H A C O M P O S I T I O N I N

S A F C T L O C A U T O M C E I L O

A C E Y B K L E T A M U R H S Y P G

S H E S P T O E R E H P S O S E M R

F O A P R G N O S P R E H V D M N A

ARGON BAROMETER COMPOSITION

HETEROSPHERE HOMOSPHERE HUMIDITY

MESOSPHERE NITROGEN OXYGEN

PSYCHROMETER TORRICELLI TROPOSPHERE

(The solution to this Word Search is at the end of Appendix E.)

# Supplemental Activity 4

## Hydrosphere Vocabulary

P T R M C F R E T A W H S E R F U K L R
X R M Y L A R S T E K F A S E C Z T T E
R O E P V I N F I L T R A T I O N S C T
K P U F T C A S R S C O T G C P R H Y A
P I L U J V K M N I M P O A A E P F U W
R C C S C U O R A S T L E Q L A E V Q D
V S H P X T H N G T O E H T G N Y C Z N
K O P A F O X S E R U L K D H G Y R Y U
H T E V S H M S D E H E E R D C Q A T O
A K D B O M Z Y Q A S Y J L L E F I N R
Y A P X O N H S U M A L M E Q P R E L G
P E R M E A B L E S E R E H P S I M E H
M L Q I T J O P F I H C W D L N D R O N
T L W U A T O R C C O O V R O Q S P C E
H N M M A T I M P E R M E A B L E R E A
A T E O S T Z Q X H E A Y I Q L K O A V
Y O I R V E O A L U A B V N P O A R N E
O T E Q R W T R A S R E T A W T L A S Q
I O K T Q U R O U C C L B G O N S T E M
I X C O U T C E B A Y V R E S S A L K E

| | | | |
|---|---|---|---|
| CHASM | SALTWATER | OCEANS | HEMISPHERE |
| DRAINAGE | CURRENT | STREAMS | INFILTRATION |
| GLACIER | EQUATOR | CYCLE | PERMEABLE |
| HYDROLOGIC | GROUNDWATER | FRESHWATER | TROPICS |
| LAKES | IMPERMEABLE | | |

(The solution to this Word Search is at the end of Appendix E.)

# Supplemental Activity 5

## Weather Vocabulary

| | | | | | | | | | | | | | | | | | | | |
|---|---|---|---|---|---|---|---|---|---|---|---|---|---|---|---|---|---|---|---|
| G | N | I | N | T | H | G | I | L | S | O | R | Q | C | R | I | N | S | T | U |
| R | I | N | E | E | Q | L | N | K | S | L | E | A | B | A | T | H | N | W | M |
| J | E | L | I | G | C | U | L | R | U | A | Z | H | I | C | V | A | R | T | O |
| Z | N | Y | X | S | P | E | A | V | B | C | U | T | M | N | E | I | S | O | C |
| H | O | A | I | L | R | N | S | T | M | A | H | R | Y | Q | S | L | E | E | D |
| S | L | T | M | Q | E | W | C | E | I | U | W | O | N | S | U | R | R | I | C |
| T | C | A | A | B | C | R | T | O | N | R | A | O | D | I | T | P | B | E | C |
| S | Y | H | A | C | I | X | G | D | O | Z | C | L | U | S | O | C | U | S | A |
| I | C | U | N | S | P | V | E | E | L | W | M | E | O | D | R | O | L | G | S |
| G | M | U | O | O | I | R | N | S | U | S | A | E | A | C | R | L | I | T | T |
| O | E | Z | B | S | T | S | F | E | M | R | Q | N | S | A | E | K | E | I | O |
| L | S | T | W | O | A | P | D | H | U | R | R | I | C | A | N | E | I | S | N |
| O | C | I | R | N | T | U | S | T | C | O | N | M | B | U | L | S | R | I | O |
| R | S | I | M | N | I | B | O | E | T | V | S | P | E | S | A | E | H | G | G |
| O | B | T | Z | J | O | M | W | F | O | U | E | N | T | N | Q | N | V | M | I |
| E | A | I | R | A | N | Z | B | M | L | S | L | C | L | O | U | D | V | O | K |
| T | O | R | R | A | D | G | P | U | O | U | N | D | B | W | C | U | M | U | L |
| E | A | K | A | T | T | O | M | C | S | G | E | T | I | R | A | F | E | V | B |
| M | O | Y | N | V | F | U | H | G | O | O | S | H | U | S | E | P | I | E | R |
| A | E | L | L | O | C | C | S | E | W | L | B | F | P | H | A | Z | R | Y | T |

CIRRUS
CUMULUS
HURRICANE
NIMBUS
SLEET
THUNDER
CLOUD
CYCLONE
LIGHTNING
PRECIPITATION
SNOW
TORNADO
CUMULONIMBUS
HAIL
METEOROLOGIST
RAIN
STRATUS

(The solution to this Word Search is at the end of Appendix E.)

# Supplemental Activity 6

## Hands-On Time: How to Make Papier-Maché

*Adult supervision required.*

### Materials

- Flour
- Water
- Newspaper strips — 6 inches by 1 inch, approximately
- Floral wire
- Balloons
- Acrylic craft paints

### Instructions

1. Using equal parts of water and flour, combine to make a paste.
2. Adjust water and flour until desired consistency is achieved.
3. Using wire, create desired form. If you want a sphere, inflate a balloon and tie.
4. Dip newspaper in flour mixture and gently slide fingers along to remove excess paste.
5. Wrap paper around form, watching out for any large bulges.
6. Continue wrapping form until there are no gaps. Smooth over any bulges and wrinkles, using smaller strips of newspaper if necessary.
7. Set your object aside for drying time. I recommend allowing at least a full day for drying. Drying time will vary according to how many layers of paper are on the form as well as how much paste.
8. Object will dry into a hard hollow form. If using a balloon for your form, stick it with a straight pin to pop the balloon.
9. Paint object as desired.

# Supplemental Activity 7

## Hands-On Time: Recipe for Play Dough

*Adult supervision required.*

### Materials

- 2 cups flour
- 1 cup salt
- 4 teaspoons cream of tartar
- 2 tablespoons vegetable oil
- 2 cups water
- Wax paper
- Food coloring. (I recommend the type found in cake-decorating sections of craft stores. It is concentrated and produces a true, rich color with very little product.)

### Instructions

1. Mix flour, salt, and cream of tartar in a large pot.
2. Add oil and water. Mix until smooth.
3. Over medium heat, stir constantly until the mixture leaves the sides of the pot.
4. Put dough on wax paper. Allow mixture to cool until you can work it with your hands.
5. Divide dough into smaller sections, thinking how many colors you'd like to have. I recommend four sections: red, blue, green, and yellow.
6. Add appropriate food coloring a small amount at a time, until the desired effect is achieved.
7. Store in zipper-style bags or airtight containers.

# Supplemental Activity Word Search Solutions

## Atmosphere Vocabulary Solution

| | | | | | | | | | | | | | | | | | |
|---|---|---|---|---|---|---|---|---|---|---|---|---|---|---|---|---|---|
| H | | | | | | | | | | | | | | | | | |
| E | | | T | R | O | P | O | S | P | H | E | R | E | | | | |
| T | | | | | | | | | | N | | R | | | T | | |
| E | | | | | | | | | E | | E | | | | O | | |
| R | | B | | | | | | G | | H | | | | | R | | |
| O | | | A | | | | Y | | P | | | | | | R | | |
| S | | | R | R | | X | | S | | | | | | | I | | |
| P | | | E | | O | | O | | | | | | | | C | | |
| H | | | T | | | M | | | | | | | | | E | | |
| E | | | E | | O | | E | | | | | | | | L | | Y |
| R | | | M | H | | | | T | | | | | | | L | T | |
| E | | | O | | | | | | E | | | | | | I | | |
| N | I | T | R | O | G | E | N | | | R | | | | D | | | |
| | | | H | | C | O | M | P | O | S | I | T | I | O | N | | N |
| | | | C | | | | | | | | | M | | | | | O |
| | | | Y | | | | | | | | U | | | | | | G |
| | | | S | | | | E | R | E | H | P | S | O | S | E | M | R |
| | | | P | | | | | | | | | | | | | | A |

# Hydrosphere Vocabulary Solution

| | T | | | | | R | E | T | A | W | H | S | E | R | F | | | | R |
|---|---|---|---|---|---|---|---|---|---|---|---|---|---|---|---|---|---|---|---|
| | R | | | | | | | | | | | | | E | C | | | | E |
| | O | | | | I | N | F | I | L | T | R | A | T | I | O | N | | | T |
| | P | | | | | | | | | | | | G | C | | | | | A |
| | I | | | | | | | | | | | O | | A | | | | | W |
| | C | | | | | | | | S | | L | | | L | | | | | D |
| | S | H | | | | | | | T | O | | | | G | | | C | | N |
| | | | A | | | | | | R | | | | | | | Y | | | U |
| | | | | S | | | | D | E | | | | | | C | | | | O |
| | | | | | M | | Y | | A | | | | | L | | | | | R |
| | | | | | | H | | | M | | | | E | | | | | | G |
| P | E | R | M | E | A | B | L | E | S | E | R | E | H | P | S | I | M | E | H |
| | | Q | | | | | | | | | | | D | | | | | O | |
| T | | | U | | | | | | | | | | R | | | S | | C | |
| | N | | | A | | I | M | P | E | R | M | E | A | B | L | E | | E | |
| | | E | | | T | | | | | | | | I | | | K | | A | |
| | | | R | | | O | | | | | | | N | | | A | | N | |
| | | | | R | | | R | | | R | E | T | A | W | T | L | A | S | |
| | | | | | U | | | | | | | | G | | | | | | |
| | | | | | | C | | | | | | | E | | | | | | |

# Weather Vocabulary Solution

| G | N | I | N | T | H | G | I | L |  |  | R |  |  |  |  |  |  |  |  |
|---|---|---|---|---|---|---|---|---|---|---|---|---|---|---|---|---|---|---|---|
|  |  |  |  |  |  |  |  |  | S |  |  | A |  |  |  | H |  |  |  |
|  | E |  |  |  |  |  |  |  | U |  |  |  | I |  |  | A |  |  |  |
|  | N |  |  |  | P |  |  |  | B |  |  | T |  | N |  | I |  |  |  |
|  | O |  |  |  | R |  |  |  | M |  | H |  |  |  |  | L |  |  |  |
|  | L |  |  |  | E |  |  |  | I | U |  |  |  | S | U | R | R | I | C |
| T | C |  |  |  | C |  |  |  | N |  |  |  |  |  |  |  |  |  |  |
| S | Y |  |  |  | I |  |  | D | O |  |  |  |  |  | O |  |  |  |  |
| I | C |  |  |  | P |  | E |  | L |  |  |  |  | D |  |  |  |  |  |
| G |  |  |  |  | I | R |  |  | U |  |  |  | A |  |  |  |  | T |  |
| O |  |  |  |  | T |  |  |  | M |  |  | N |  |  |  |  | E |  |  |
| L |  |  |  |  | A |  |  | H | U | R | R | I | C | A | N | E |  |  |  |
| O |  |  |  | N | T |  |  |  | C | O |  |  |  |  | L |  |  |  |  |
| R | S |  |  |  | I |  |  |  | T |  | S |  |  | S |  |  |  |  |  |
| O |  | T |  |  | O | M |  |  |  | U |  |  |  | N |  |  |  |  |  |
| E |  |  | R |  | N |  | B |  | L |  |  | C | L | O | U | D |  |  |  |
| T |  |  |  | A |  |  |  | U |  |  |  |  |  | W |  |  |  |  |  |
| E |  |  |  |  | T |  | M |  | S |  |  |  |  |  |  |  |  |  |  |
| M |  |  |  |  |  | U |  |  |  |  |  |  |  |  |  |  |  |  |  |
|  |  |  |  |  | C |  | S |  |  |  |  |  |  |  |  |  |  |  |  |

## Appendix F

# HOW TO MAKE A FOLDERBOOK

Making a folderbook is a simple way to tie together what you're learning in a certain area. You can create a folderbook on any subject.

A folderbook is as simple as a single file folder with pieces of related learning attached inside. Use your imagination and creativity to make your folderbook unique as well as informative. Work with colorful inks and papers.

Here are some more suggestions, but don't be limited by these.

- Include pictures and/or maps.
- Paste colorful envelopes inside and put cards with information or smaller books inside the envelopes.
- Make small books or flipbooks and paste them inside.
- Decorate lists and glue them in.
- Don't forget to use the back of the folder to display more information.
- Decorate the front cover in an appropriate fashion.

- Somewhere in the folderbook, be sure to include your name and the date you finished the folderbook!

*Example:* A folderbook on volcanoes could include:

- A map or cutaway illustration of the inside of a volcano
- An envelope containing index cards with important facts about several well-known volcanoes and/or about different types of volcanoes
- An index card with volcano-related vocabulary words listed and defined
- A flipbook illustrating a volcanic eruption
- Brief excerpts from firsthand accounts of people who have survived volcanic eruptions
- A clipping from a newspaper or magazine article about volcanoes or about a recent volcanic event
- A map showing the location of the famous volcanoes you wrote about

Here are some pictures of a volcano folderbook to help spark your creativity!

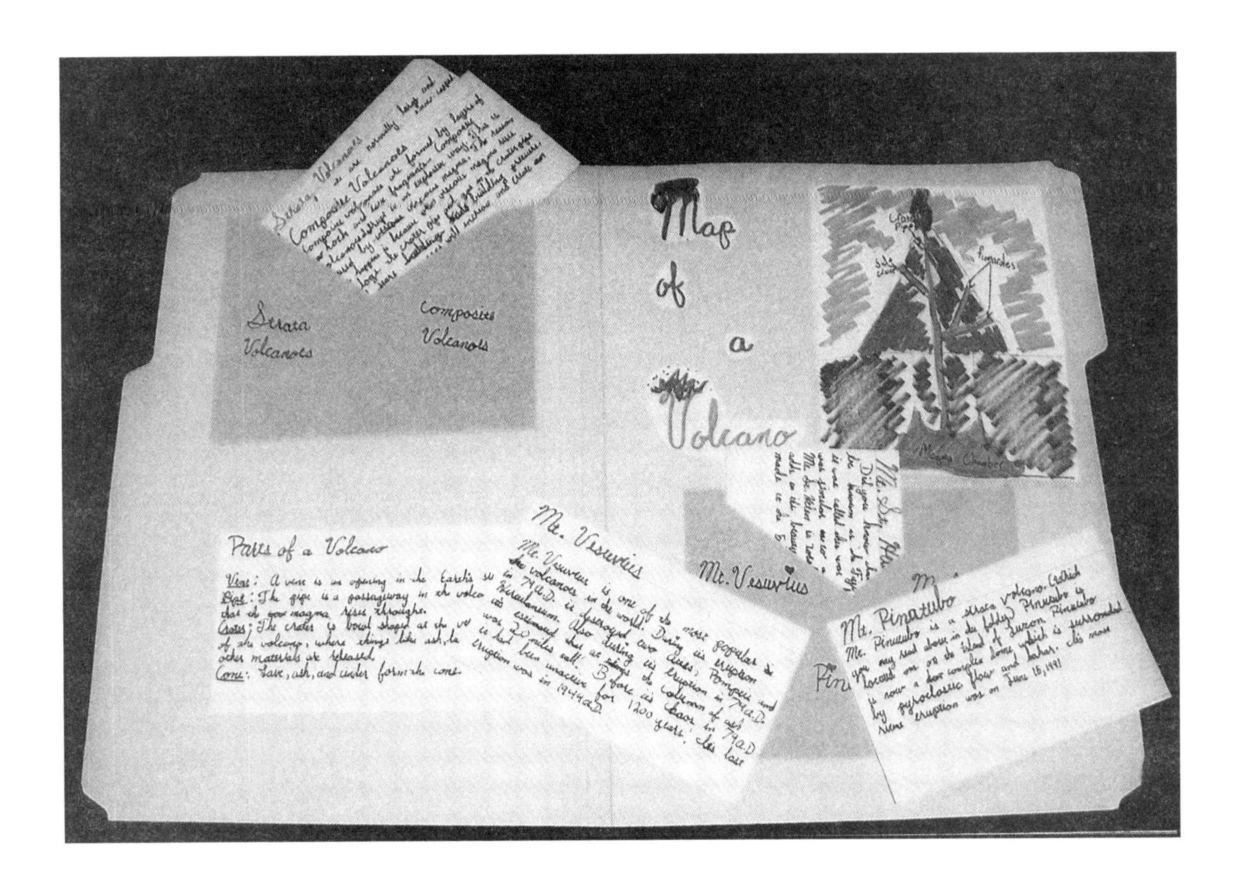
Composite Volcanoes
Map of a Volcano
Magma Chamber
Parts of a Volcano
Mt. Vesuvius
Mt. Pinatubo

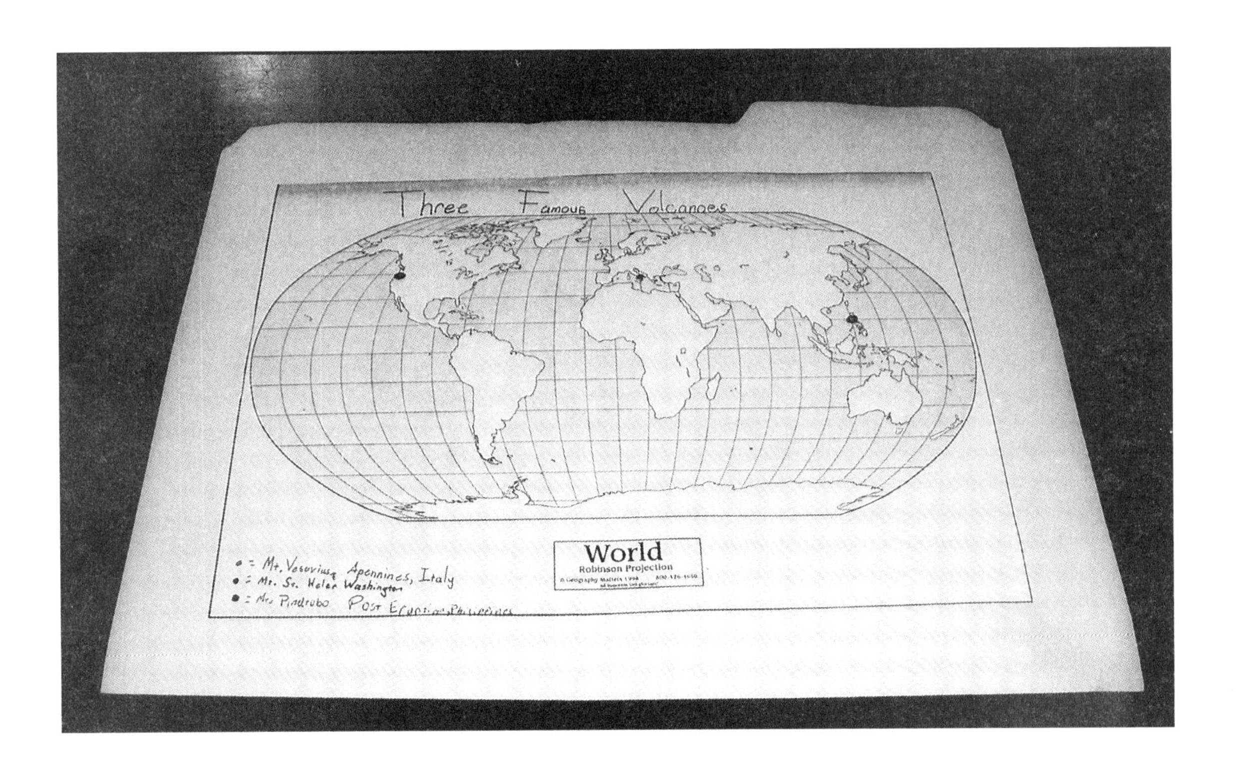
Three Famous Volcanoes
World
Robinson Projection
Apennines, Italy

Appendix G

# BOOK AND RESOURCE LIST

*Note that entries preceded by a cross are Christian in content.*

## Unit One: Getting Started

† *The Amazing Story of Creation: From Science and the Bible*, Duane T. Gish, D. Dish, Earl Snellenberger (Illustrator), Bonnie Snellenberger (Illustrator). (Master Books; ISBN: 0890511209; 1996).

*Dr. Art's Guide to Planet Earth: For Earthlings Ages 12 to 120*, Art Sussman, Ph.D., Emiko Koike (Illustrator). (Chelsea Green Publishing Company; ISBN: 189013273X; 2000). 128 pp. Grades 4–6.

*The Earth Science Book: Activities for Kids*, Dinah Zike, Jessie J. Flores. (Jossey-Bass; ISBN: 0471571660; 1993). 128 pp. Grades 4–6.

*Encyclopedia of Planet Earth*, Anna Claybourne, Gillian Doherty, Rebecca Treays, Laura Fearn, Melissa Alaverdy. (Usborne Books; ISBN: 0794505341; 2003). 160 pp. All Ages.

*Essential Atlas of Physical Geography*, Parramon Studios. (Barron's Educational Series; ISBN: 0764125117; 2003). 96 pp. All Ages.

† *Exploring Planet Earth*, John Hudson Tiner. (Master Books; ISBN: 0890511780; 1997). 160 pp. Grade 5 and Up.

*Forces of Nature : The Awesome Power of Volcanoes, Earthquakes, and Tornadoes*, Catherine O'Neill Grace. (National Geographic Children's Books; ISBN: 0792263286; 2004). 64 pp. Young Adult.

*What Is the World Made Of? All About Solids, Liquids, and Gases* (Let's-Read-and-Find-Out Science, Stage 2), Kathleen Weidner Zoehfeld, Paul Meisel. (HarperTrophy; ISBN: 0064451631; 1998). 32 pp. Grades 4–8.

## Unit Two: The Lithosphere

*Caves and Caverns*, Gail Gibbons. (Voyager Books; ISBN: 0152013652; 1996). 32 pp. Grades 1–5.

*Cave Sleuths: Solving Science Underground* (Science On The Edge), Laurie Lindop. (21st Century; ISBN: 0761327029; 2004). 80 pp. Grades 4–6.

*Earth* (Eyewitness Science), Susanna VanRose. (DK Publishing, Inc.; ISBN: 0789455757; 2000). 64 pp. Grades 1–6.

*Earthquake!*, Cynthia Pratt Nicolson. (Kids Can Press; ISBN: 1550749684; 2002). 32 pp. Grades 1–5.

*Earthquake Emergency* (Expedition Earth), Dougal Dixon, Neil Morris. (Waterbird Press; ISBN: 1577688481; 2004). 32 pp. Grades 1–5.

*Earthquakes* (Earth Science), Trudi Strain Trueit. (Franklin Watts; ISBN: 0531162435; 2003). 64 pp. Grades 1–5.

*Earthquakes* (Disasters in Nature), Catherine Chambers. (Heinemann Library; ISBN: 1588103331; 2001). 48 pp. Grades 4–6.

*Earthquakes* (True Books-Nature), Paul P. Sipiera. (Children's Press; ISBN: 051626432X; 1999). 48 pp. Grades 4–6.

*Earthquakes*, Keith Lye. (Steck-Vaughn; ISBN: 0811496570; 1996). 32 pp. All Ages.

*Erosion: How Land Forms, How It Changes* (Exploring Science), Darlene R. Stille. (Compass Point Books; ISBN: 0756508541; 2005). 48 pp. Grades 4–6.

*Eyewitness: Rocks and Minerals*, R. F. Symes. (DK Publishing, Inc.; ISBN: 0789458047; 2000). 64 pp. Ages 9–12.

*Forest Fire* (Expedition Earth), Dougal Dixon, Neil Morris. (Waterbird Press; ISBN: 157768849X; 2004). 32 pp. Grades K–5.

† *The Fossil Book* (Wonders of Creation), Gary Parker, Mary Parker. (Master Books; ISBN: 0890514380; 2006). 80 pp. Junior High–Adult.

*Furious Earth: The Science and Nature of Earthquakes, Volcanoes, and Tsunamis*, Ellen J. Prager. (McGraw-Hill; ISBN: 0071351612; 1999). 235 pp. Grades 6–Adult.

*Gem Hunter's Kit*, Belinda Recio. (Running Press Book Publishers; ISBN: 0762412127; 2002). 64 pp. Grades 4–6.

† *Geology Book*, John D. Morris. (Master Books; ISBN: 0890512817; 2000). 80 pp. All Ages.

*Geology Rocks! 50 Hands-On Activities to Explore the Earth*, Cindy Blobaum, Michael P. Kline. (Williamson Publishing Company; ISBN: 1885593295; 1999). 96 pp. Grades 4–6.

† *Grand Canyon: Monument to Catastrophe*, Austin. (Institute for Creation Research; ISBN: 0932766331; 1995). Adult.

*The Magic School Bus Blows Its Top: A Book About Volcanoes* (Magic School Bus), Gail Herman, Bob Ostrom. (Scholastic Paperbacks; ISBN: 0590508350; 1996). 32 pp. Grades K–6.

*The Magic School Bus Inside the Earth* (Magic School Bus), Joanna Cole. (Scholastic; ISBN: 0590407600; reprint 1989). 40 pp. Grades K–6.

*Plate Tectonics* (Great Ideas of Science), Rebecca L. Johnson. (Lerner Publications; ISBN: 0822530562; 2005). 80 pp. Grades 4–6.

*The Rock Factory: A Story About the Rock Cycle* (Science Works), Jacqui Bailey, Matthew Lilly (Illustrator). (Picture Window Books; ISBN: 1404815961; 2006). 32 pp. Ages 4–8.

*Rocks and Fossils* (Usborne), Martyn Bramwell. (E.D.C. Publishing; ISBN: 0746019750; 1997). 32 pp. Grades 4–6.

*Rocks and Minerals* (Eyewitness Explorers), Steve Parker. (DK Publishing, Inc.; ISBN: 0789416824; 1997). 64 pp. Ages 4–8.

*Rocks, Gems and Minerals* (Earth Science), Trudi Strain Trueit. (Franklin Watts; ISBN: 0531162419; 2003). 64 pp. Grades K–5.

*Shattering Earthquakes* (Awesome Forces of Nature), Louise Spilsbury, Richard Spilsbury. (Heinemann; ISBN: 1403454469; 2004). 32 pp. Grades K–6.

*Time For Kids: Volcanoes!* (Time For Kids). (HarperCollins; ISBN: 0060782242; 2006). 32 pp. Grades 4–6.

*Thundering Landslides* (Awesome Forces of Nature), Louise Spilsbury, Richard Spilsbury. (Heinemann; ISBN: 1403454442; 2004). 32 pp. Grades K–6.

*Violent Volcanoes* (Awesome Forces of Nature), Louise Spilsbury, Richard Spilsbury. (Heinemann; ISBN: 1403454485; 2004). 32 pp. Grades K–6.

*Volcano!*, Cynthia Pratt Nicolson. (Kids Can Press; ISBN: 1550749668; 2001). 32 pp. Grades K–6.

*Volcano Evacuation* (Expedition Earth), Dougal Dixon, Neil Morris. (Waterbird Press; ISBN: 1577688694; 2004). 32 pp. Grades K–6.

*Volcanoes* (True Books-Nature), Paul P. Sipiera. (Children's Press; ISBN: 0516264443; 1999). 48 pp. Grades 4–6.

*Volcanoes* (Earth Science), Trudi Strain Trueit. (Franklin Watts; ISBN: 0531162443; 2003). 64 pp. Grades K–6.

*Volcanoes* (WorldLife Library Series), Peter Clarkson. (Voyageur Press; ISBN: 0896585026; 2000). 72 pp. All Ages.

*Volcanoes* (Disasters in Nature), Catherine Chambers. (Heinemann Library; ISBN: 1588103366; 2001). 48 pp. Grades 4–6.

*Volcanoes and Earthquakes* (Eyewitness Books). (DK Publishing, Inc.; ISBN: 0756607353; 2004). 72 pp. Grades K–6.

*Why Do Volcanoes Blow Their Tops?: Questions and Answers About Volcanoes and Earthquakes* (Scholastic Q & A), Melvin Berger, Gilda Berger, Higgins Bond (Illustrator). (Scholastic Reference; ISBN: 0439095808; 2000). 48 pp. Ages 9–12.

*Wildfires,* Seymour Simon. (HarperTrophy; ISBN: 0688175309; 2000). 32 pp. Ages 4–8.

*Wildfires* (True Books-Nature), Paul P. Sipiera, Diane M. Sipiera. (Children's Press; ISBN: 0516264451; 1999). 47 pp. Grades 4–6.

## Related Literature

*The Adventures of Tom Sawyer,* Mark Twain. (Penguin Classics; ISBN: 0140390839; 1986). 256 pp. All Ages.

*Dear Katie, the Volcano Is a Girl,* Jean Craighead George. (Hyperion; ISBN: 0786803142; 1998). 32 pp. Grades 4–6.

*Journey to the Centre of the Earth,* Jules Verne. (Penguin; ISBN: 0140022651; 1965). 256 pp.

*Julie the Rockhound,* Gail Langer Karwoski, Chad Wallace (Illustrator). (Sylvan Dell Publishing; ISBN: 097649437X; 2006). 32 pp. Grades 4–6.

*Red Sails to Capri,* Ann Weil. (Puffin Books; ISBN: 0140328580; 1989). 160 pp. Grades 4–6.

*Rocky Road Trip* (Magic School Bus Chapter Book #20): Rocks and Minerals (Magic School Bus), Judith Stamper, Hope Gangloff (Illustrator). (Scholastic Paperbacks; ISBN: 0439560535; 2004). 96 pp. Grades 3–5.

## Unit Three: The Hydrosphere

*Awesome Ocean Science!*, Cindy A. Littlefield, Sarah Rakitin. (Williamson Publishing Company; ISBN: 1885593716). 120 pp. Grades 4–6.

*Big Freeze* (Wild Weather), Catherine Chambers. (Heinemann; ISBN: 1403401098; 2002). 32 pp. Grades K–4.

*Bill Nye the Science Guy's Big Blue Ocean*, Bill Nye. (Hyperion; ISBN: 0786842210; 1999). 48 pp. Grades 4–6.

*Crushing Avalanches* (Awesome Forces of Nature), Louise Spilsbury, Richard Spilsbury. (Heinemann; ISBN: 1403442304; 2003). 32 pp. Grades K–6.

*Dreadful Droughts* (Awesome Forces of Nature), Louise Spilsbury, Richard Spilsbury. (Heinemann; ISBN: 1403442312; 2003). 32 pp. Grades K–6.

*Drought* (Wild Weather), Catherine Chambers. (Heinemann; ISBN: 140340111X; 2002). 32 pp. Grades K–3.

*Flood* (Wild Weather), Catherine Chambers. (Heinemann; ISBN: 1403401128; 2002). 32 pp. Grades 4–6.

*Floods!*, Lorraine Jean Hopping, Jody Wheeler. (Scholastic; ISBN: 0439087570; 2000). 48 pp. Grades K–3.

*Floods* (True Books-Nature), Paul P. Sipiera, Diane M. Sipiera. (Children's Press; ISBN: 0516264346; 1999). 47 pp. Grades 4–6.

*I Wonder Why the Sea Is Salty: And Other Questions About the Oceans* (I Wonder Why), Anita Ganeri. (Kingfisher; ISBN: 0753456117; 2003). 32 pp. Ages 4–8.

*The Magic School Bus On the Ocean Floor* (Magic School Bus), Joanna Cole, Bruce Degen (Illustrator). (Scholastic Press; ISBN: 0590414313; 1994). 48 pp. Grades K–6.

*The Magic School Bus Wet All Over: A Book About the Water Cycle* (Magic School Bus), Pat Relf, Carolyn Bracken. (Scholastic Press; ISBN: 0590508334; 1996). 32 pp. Grades K–6.

*Ocean* (Eyewitness Books). (DK Publishing, Inc.; ISBN: 0756607116; 2004). 72 pp. Grades K–6.

† *Ocean Book* (Wonders of Creation), Frank Sherwin, Bryan Miller, Debbie Brooks, Beth Wiles. (Master Books; ISBN: 0890514011; 2004). 80 pp. Grades 4–8.

*Oceans: An Activity Guide for Ages 6–9*, Nancy F. Castaldo. (Chicago Review Press; ISBN: 1556524439; 2002). 134 pp. Grades K–3.

*Pond and River* (Eyewitness Books). (DK Children; ISBN: 0756610869; 2004). 72 pp. Grades K–6.

*Raging Floods* (Awesome Forces of Nature), Louise Spilsbury, Richard Spilsbury. (Heinemann; ISBN: 1403442320; 2003). 32 pp. Grades K–6.

*Seashore* (Eyewitness Books). (DK Children; ISBN: 0756607213; 2004). 72 pp. Grades K–6.

*Sweeping Tsunamis* (Awesome Forces of Nature), Louise Spilsbury, Richard Spilsbury. (Heinemann; ISBN: 1403442339; 2003). 32 pp. Grades K–6.

*Tsunami Man:* Learning About Killer Waves with Walter Dudley, Anthony D. Fredericks. (University of Hawaii Press; ISBN: 0824824962; 2002). 79 pp. Grades 4–6.

*Water Cycle* (Earth Science), Trudi Strain Trueit. (Franklin Watts; ISBN: 0531162206; 2002). 63 pp. K–6.

## Related Literature

*Carry On, Mr. Bowditch*, Jean Lee Latham. (Houghton Mifflin; ISBN: 0618250743; 2003). 256 pp. Grade 5–Adult.

*Dove*, Robin L. Graham. (Harper Paperbacks; ISBN: 0060920475; 1991). 240 pp. Young Adult.

*Paddle-to-the-Sea*, Holling C. Holling. (Houghton Mifflin; ISBN: 0395150825; 1941). 64 pp. Ages 4–10.

*Stowaway*, Karen Hesse. (First Aladdin Paperbacks; ISBN: 0689839898; 2002). 320 pp. Ages 9–12.

*The Terrible Wave*, Marden Dahlstedt. (Putnam Publishing Group; ISBN: 0698201884; 1972). 125 pp. Grade 4 and Up.

*20,000 Leagues Under the Sea*, Jules Verne. (Tor Classics; ISBN: 0812550927; 1995). 320 pp. Grade 4 and Up.

*The Voyager's Stone: The Adventures of a Message-Carrying Bottle Adrift on the Ocean Sea*, Robert Kraske. (Scholastic; ISBN: 0531068900; 1995). Grades 4–6.

## Unit Four: The Atmosphere

*Air* (Against the Elements), Jane Walker. (Copper Beech; ISBN: 0761308555; 1998). 32 pp. Grades 4–6.

*Air* (Interfact book and disk). (Two-Can Publishing; ISBN: 1587284545; 2000). 48 pp. Grades K–3.

*Air* (Picture Science). (Franklin Watts; ISBN: 0531142019; 1992). 29 pp. Ages 4–8.

*Air, Air All Around*, Joanne Barkan. (Silver Press; ISBN: 0671686593; 1990). 30 pp. Grades K–2.

*Air Is All Around You*, Franklyn M. Branley. (HarperTrophy; ISBN: 0064450481; 1986). 32 pp. Grades K–3.

*Air, Water and Weather* (Discovering Science), Michael Pollard. (BLA Publishing; 1987). 47 pp. Grades 1–6.

*The Atmosphere* (Hands-On Science), John Owen Edward Clark. (Franklin Watts; ISBN: 0531173674; 1992). 32 pp. Grades 4–6.

*Atmosphere and Weather* (Hands-On Science), Karen Kwitter. (Walch; ISBN: 0825137640; 1998). Ages 4–8.

*Atmosphere: Sea of Air,* Roy A. Gallant. (Benchmark Books, NY; ISBN: 0761413669; 2002). 80 pp. Ages 9–12.

*Experiments with Air* (True Books-Science Experiments), Salvatore Tocci. (Children's Press; ISBN: 0516293621; 2003). 48 pp. Grades 3–5.

*Exploring the Sky by Day,* Terrence Dickinson. (Firefly Books Ltd; ISBN: 0920656714; 1988). 72 pp. Grades 5–Adult.

*A Field Guide to the Atmosphere* (Peterson Field Guides), John A. Day, Vincent J. Schaefer, Christy E. Day (Illustrator). (Houghton Mifflin; ISBN: 0395976316; 1998). 380 pp. Adult.

*Hot Air: The (Mostly) True Story of the First Hot-Air Balloon Ride* (Caldecott Honor Book), Marjorie Priceman. (Atheneum/Anne Schwartz Books; ISBN: 0689826427; 2005). Grades 4–6.

*Lightning and Other Wonders of the Sky* (Amazing Science), Q. L. Pearce, Mary Ann Fraser (Illustrator). (Julian Messner; ISBN: 0671686488; 1989). 64 pp.

*Looking at the Sky* (My First Field Guides), Jennifer Frantz. (Grosset & Dunlap; ISBN: 0448424886; 2002). 64 pp. Ages 4–8.

† *Over Our Heads In Wonder: Science and the Sky,* Esther Yant. (Bright Ideas Press; ISBN: 1892427028; 2001). 84 pp. Grades K–5.

*The Science of Air: Projects and Experiments With Air and Flight* (Tabletop Scientist), Steve Parker. (Heinemann; ISBN: 1403472807; 2005). 32 pp. Grades 4–6.

*The Sky's the Limit,* Mark Rauzon. (Millbrook Press; ISBN: 0761312633; 1999). 32 pp. Grades 4–6.

*Temperature,* Alan Rodgers, Angella Streluk. (Heinemann; ISBN: 1403401292; 2002). 32 pp. Grades 4–6.

*Weather.* (Reader's Digest; ISBN: 0895779757; 1987). 155 pp. Ages 10–Adult.

*What's Up with Altitude: Mr. Moffat's Class Investigates How Altitude Affects Our Bodies* (CMC Wilderness Kids Series), Lisa Gardiner. (Colorado Mountain Club Press; ISBN: 0972441387; 2004). 45 pp. All Ages.

### Related Literature

† *My Angel Named Herman,* Elmer Towns. (Tommy Nelson; ISBN: 0849958393; 1998). 36 pp. Ages 4–8.

*Snowflake Bentley* (Caldecott Medal Book), Jacqueline Briggs Martin, Mary Azarian (Illustrator). (Houghton Mifflin; ISBN: 0395861624; 1998). 32 pp. Ages 4–8.

*The Twenty-One Balloons,* William Pene du Bois. (Puffin Books; ISBN: 0140320970; 1975). 180 pp. Ages 9–12.

## Unit Five: Earth's Weather

*Arty Facts: Weather and Art Activities,* Janet Sacks. (Crabtree Publishing Company; ISBN: 0778711463; 2002). 48 pp. Ages PreK–Adult.

*Big Freeze* (Wild Weather), Catherine Chambers. (Heinemann; ISBN: 1403401098; 2002). 32 pp. Grades K–2.

*Blizzard* (Wild Weather), Catherine Chambers. (Heinemann; ISBN: 1403401101; 2002). 32 pp. Grades K–2.

*Cloud Book,* Tomie De Paola. (Holiday House; ISBN: 0823405311; 1984). 32 pp. Grades K–3.

Cloud Mobile Kit, Jim Green. (Williwaw Publishing Co.; ISBN: 1892337274; 2003). Grades PreK–Adult.

*Clouds,* Gail Sounders-Smith. (Capstone Press; ISBN: 1560657774; 1998). 24 pp. Grades K–3.

*Drought* (Wild Weather), Catherine Chambers. (Heinemann; ISBN: 140340111X; 2002). 32 pp. Grades K–2.

*Exploring the Sky by Day,* Terrence Dickinson. (Firefly Books, Ltd; ISBN: 0920656714; 1988). 72 pp. Grades 5–Adult.

*Extraordinary Wild Weather,* Paul Dowswell. (Scholastic; ISBN: 0439286026; 2003). 32 pp. Grades 1–4.

*Extreme Weather: A Guide and Record Book,* Christopher C. Burt. (W.W. Norton & Company; ISBN: 0393326586; 2004). 304 pp.

*Flood* (Wild Weather), Catherine Chambers. (Heinemann; ISBN: 1403401128; 2002). 32 pp. Grades 2–6.

*Heat Wave* (Wild Weather), Catherine Chambers. (Heinemann; ISBN: 1403401136; 2002). 32 pp. Grades K–2.

*Hurricane* (Wild Weather), Catherine Chambers. (Heinemann; ISBN: 1403401144; 2002). 32 pp. Grades K–2.

*Hurricane and Tornado* (Eyewitness Books), Jack Challoner. (DK Publishing, Inc.; ISBN: 075660690X; 2004). 72 pp. Grades 4–8.

*The Kids' Book of Weather Forecasting,* Mark Breen, Kathleen Friestad, Michael Kline. (Williamson Publishing Company; ISBN: 1885593392; 2000). 144 pp. Grades 2–8.

*Lightning,* Gail Saunders-Smith. (Capstone Press; ISBN: 1560657790; 1998). 24 pp. Grades K–3.

*Lightning and Other Wonders of the Sky,* Q. L. Pearce. (Julian Messner; ISBN: 0671686488; 1989). 64 pp. Ages 9–12.

*The Magic School Bus Inside a Hurricane* (Magic School Bus), Joanna Cole. (Scholastic Press; ISBN: 0590446878; 1996). 48 pp. Ages 9–12.

*My First Pocket Guide: Weather*, Patricia Daniels. (National Geographic; ISBN: 0792265882; 2001). 80 pp. Grades 1–5.

*Restless Skies: The Ultimate Weather Book*, Paul Douglas. (Sterling; ISBN: 0760761132; 2004). 256 pp.

*Scholastic Atlas of Weather*. (Scholastic Reference; ISBN: 0439419026; 2004). 80 pp. Grades 3–7.

*Seasons*, Robyn Supraner. (Troll Communications; ISBN: 08167471990; 2003). 46 pp. Grades K–3.

*Seasons and Weather* (Let's Explore Science), David Evans, Claudette Williams. (DK Publishing, Inc.; ISBN: 154582094; 1993). 29 pp. Grades K–3.

*Sunshine Makes the Seasons* (Let's-Read-and-Find-Out Science 2), Franklyn M. Branley, Giulio Maestro (Illustrator). (HarperTrophy; ISBN: 0060592052; 2005). 40 pp. Ages 9–12.

*Thunderstorm* (Wild Weather), Catherine Chambers. (Heinemann; ISBN: 1403401152; 2002). pp. 32. Grades 2–6.

*Tornado* (Wild Weather), Catherine Chambers. (Heinemann; ISBN: 1403401160; 2002). 32 pp. Grades K–2.

*Weather* (Eyewitness Explorers), John Farndon. (DK Publishing, Inc.; ISBN: 0789429853; 1998). 61 pp. Ages 4–8.

*Weather*, Janice Van Cleave. (Wiley; ISBN: 047103231X; 1995). 96 pp. Grades 3–7.

† *Weather* (Made by God). (In Celebration; ISBN: 0742428095). 32 pp. Grades K–3.

*Weather*. (Reader's Digest; ISBN: 0895779757; 1987). 155 pp. Ages 10–Adult.

*Weather*, Seymour Simon. (HarperTrophy; ISBN: 068817521X; 2000). 40 pp. K–Adult.

*Weather* (Super Science Activities), Ruth Young. (Teacher Created Resources; ISBN: 0743936671; 2002). 48 pp. Grades 2–5.

*Weather* (Topic Books), Fiona MacDonald. (Franklin Watts; ISBN: 0531154289; 2000). 32 pp. Grades 2–5.

*Weather*, Valerie Wyatt, Brian Share. (Kids Can Press; ISBN: 1550748157; 2000). Grades 3–7.

*Weather and Climate*, F. Watt. (E.D.C. Publishing; ISBN: 0746006837; 1991). 48 pp. Grades 5–10.

*Weather and Climate* (Young Discoverers), Barbara Taylor. (Kingfisher; ISBN: 0753455099; 2002). 32 pp. Grades K–3.

† *Weather Book*, Michael J. Oard. (Master Books; ISBN: 0890512116; 1997). 80 pp. Grades 5–Adult.

*Weather Dude: A Musical Guide to the Atmosphere* (Audiocassette), Nick Walker. (Small Gate Media; ISBN: 0964338904; 1995). 46 pp.

*Weather Eyewitness*, Brian Cosgrove. (DK Publishing, Inc.; ISBN: 0789457822; 2000). 64 pp. Grades K–3.

*Weather Identification Handbook*, Storm Dunlop. (Lyons Press; ISBN: 1585748579; 2003). 192 pp. Grades 5–12.

*Weather Wizard's Cloud Book*, Lewis D. Rubin. Grades PreK–Adult.

*Weather Words and What They Mean*, Gail Gibbons. (ISBN: 0823408051). 32 pp. Ages 4–8.

*Wild Weather Cards* (Games for Your Brain), Tina L. Seelig. (Chronicle Books; ISBN: 0811828824; 2000). 68 pp. Grades K–6.

## Related Literature

*Cloudy With a Chance of Meatballs*, Judi Barrett, Ron Barrett (Illustrator). (Aladdin; ISBN: 0689707495; 1982). 32 pp. Ages 4–8.

*The Wizard of Oz*, L Frank Baum. (Tor Classics; ISBN: 0812523350; 1993). 256 pp. All Ages.

# Unit Six: Beyond Earth

*Asteroids, Comets and Meteors*, Giles Sparrow. (Heinemann; ISBN: 1588109593; 2002). 40 pp. Grades 4–6.

† *Astronomy and the Bible*, Donald B. DeYoung. (Baker Book House; ISBN: 0801029910; 1992). 144 pp. Grades 9–Adult.

† *Astronomy Book*, Jonathan Henry. (Master Books; ISBN: 0890512507; 1999). 80 pp. Grades 5–Adult.

*Astronomy for All Ages*, Philip Harrington, Edward Pascuzzi. (Globe Pequot; ISBN: 0762708093; 2000). 224 pp. Grades PreK–Adult.

*Comets and Meteor Showers* (True Books-Space), Paul P. Sipiera. (Children's Press; ISBN: 0516261665; 1997). 48 pp. Grades 3–5.

*Complete Book of Our Solar System*, Vincent Douglas. (American Education Publishing; ISBN: 1577686055; 2002). 352 pp. Grades 1–3.

*Find the Constellations*, H.A. Rey. (Houghton Mifflin; ISBN: 0395244188; 1976). 80 pp. Grades K–6.

*First Guide to the Universe*, Sheila Snowder, Jane Chisholm, Lynn Myring. (Educational Development Corporation; ISBN: 0860206114; 1983). 72 pp. Grades K–4.

*Glow-in-the-Dark Constellations: A Field Guide*, C. E. Thompson. (Grosset & Dunlap; ISBN: 0448412535; 1999). 32 pp. Grades PreK–Adult.

† *God Created the World and the Universe*, Earl Snellenberger, Bonnie Snellenberger. (Master Books; ISBN: 0890511497; 1989). 32 pp. Ages 4–10.

*How the Universe Works* (How It Works), Heather Couper, Nigel Henbest. (Reader's Digest; ISBN: 089577576X; 1994). 160 pp. Ages 9–12.

*Jupiter* (True Books-Space), Larry Dane Brimner. (Children's Press; ISBN: 0516264958; 1999). 48 pp. Grades 3–5.

*Jupiter* (Exploring the Solar System), Giles Sparrow. (Heinemann; ISBN: 1588109607; 2002). 40 pp. Grades 4–6.

*The Magic School Bus Lost in the Solar System* (Magic School Bus), Joanna Cole. (Scholastic Press; ISBN: 0590414291; 1992). 40 pp. Ages 4–8.

*Mars* (True Books-Space), Larry Dane Brimner. (Children's Press; ISBN: 0516264354; 1999). 47 pp. Grades 3–5.

*Mars* (Exploring the Solar System), Giles Sparrow. (Heinemann; ISBN: 1588109615; 2002). 40 pp. Grades 4–6.

*Mars* (Eyewitness Books), Stuart Murray. (DK Publishing, Inc.; ISBN: 0756607855; 2004). 72 pp. Grades 3–6.

*Mercury* (True Books-Space), Larry Dane Brimner. (Children's Press; ISBN: 0516264362; 1999). 48 pp. Grades 3–5.

*Mercury* (Exploring the Solar System), Giles Sparrow. (Heinemann; ISBN: 1588109623; 2002). 40 pp. Grades 4–6.

*Moon* (Exploring the Solar System), Giles Sparrow. (Heinemann; ISBN: 1588109631; 2002). 40 pp. Grades 4–6.

*Moons* (True Books-Space), Allison Lassieur. (Children's Press; ISBN: 0516271865; 2000). 48 pp. Grades 3–5.

*Neptune* (True Books-Space), Larry Dane Brimner. (Children's Press; ISBN: 0516264966; 1999). 47 pp. Grades 3–5.

✝ *Our Created Moon: Earth's Fascinating Neighbor*, Donald B. DeYoung, John Whitcomb. (Master Books; ISBN: 0890514038; 2003). 160 pp. Grades 6–Adult.

*Pluto* (True Books-Space), Larry Dane Brimner. (Children's Press; ISBN: 0516264990; 1999). 48 pp. Grades 3–5.

*Saturn* (True Books-Space), Larry Dane Brimner. (Children's Press; ISBN: 0516265016; 1999). 48 pp. Grades 3–5.

*Saturn* (Exploring the Solar System), Giles Sparrow. (Heinemann; ISBN: 158810964X; 2002). 40 pp. Grades 4–6.

*So That's How the Moon Changes Shape,* Allan Fowler. (Children's Press; ISBN: 0516449176; 1991). 31 pp. Ages 4–8.

*The Solar System* (True Books-Space), Paul P. Sipiera. (Children's Press; ISBN: 0516261754; 1997). 48 pp. Grades 3–5.

*Space* (Worldwise), Carole Stott. (Franklin Watts; ISBN: 0531152693; 1995). 39 pp. Grades K–4.

*The Stargazer's Guide to the Galaxy,* Q. L. Pearce. (Torkids; ISBN: 0812594231; 1991). 64 pp. Grades PreK–Adult.

*Stargazers,* Gail Gibbons. (Holiday House; ISBN: 0823415074; 1999). Grades K–3.

† *Starlight and Time: Solving the Puzzle of Distant Starlight in a Young Universe,* D. Russell Humphreys. (Master Books; ISBN 0890512027; 1994). 137 pp. Adult.

*Stars and Planets* (Eyewitness Handbooks), Ian Ridpath. (DK Publishing, Inc., ISBN: 0789435217; 1998). 224 pp. Grades 6–Adult.

*Starwatch: Month-by-Month Guide to the Night,* Robin Kerrod. (Barron's Educational Series; ISBN: 0764156667; 2003). 96 pp. Grades PreK–Adult.

*Sun* (Exploring the Solar System), Giles Sparrow. (Heinemann; ISBN: 1588109658; 2002). 40 pp. Grades 4–6.

*The Sun,* Seymour Simon. (HarperTrophy; ISBN: 0688092365; 1989). 32 pp. Ages 4–8.

*The Sun Is Always Shining Somewhere,* Allan Fowler. (Children's Press; ISBN: 0516449060; 1992). Ages 4–8.

*Universe* (Eyewitness Books). (DK Publishing, Inc.; ISBN 0789492385; 2003). 64 pp. Ages 9–12.

*Uranus, Neptune and Pluto* (Exploring the Solar System), Giles Sparrow. (Heinemann; ISBN: 1588109666; 2002). 40 pp. Grades 4–6.

*Venus* (Exploring the Solar System), Giles Sparrow. (Heinemann; ISBN: 1588109674; 2002). 40 pp. Grades 4–6.

*Venus,* Steven Kipp. (Capstone Press; ISBN: 1560656093; 1998). 24 pp. Grades 1–4.

## Related Fiction

† *Out of the Silent Planet,* C. S. Lewis. (Scribner; ISBN: 0743234901; 2003). 160 pp.

† *Perlandra,* C. S. Lewis. (Scribner; ISBN: 07432349X; 2003). 192 pp.

† *That Hideous Strength,* C. S. Lewis. (Scribner; ISBN: 0743234928; 2003). 384 pp.

*A Wrinkle in Time,* Madeline L'Engle. (Yearling; ISBN: 0440498058; 1973). 256 pp. Grade 5 and Up.

# INDEX

Page numbers in *italics* indicate illustrations

## C



## D



## E

## R



## S

### W

# ABOUT THE AUTHOR

As an Air Force officer's wife, mother of three, Bible study teacher, ministry team leader and author, Stephanie Redmond appreciates the demands on today's woman. Still, homeschooling her children remains a priority over other commitments. As one who struggled with science, she seeks to make her curriculum user friendly, both to the students and the teachers. Her joys are the Lord, her family, singing, and most recently exploring her new home of Hawaii. Stephanie and her husband, Andy, have been married since 1983 and are the proud parents of Mike (19), Taylor (16) and Rachel (14).

# Also Available from Bright Ideas Press...

## The Mystery of History Volumes I & II by Linda Hobar

This award-winning series provides a historically accurate, Bible-centered approach to learning history. The completely chronological lessons shed new light on who walked the earth when, as well as on where important Bible figures fit into secular history. Grades 4 – 8, yet easily adaptable.

- Volume I: Creation to the Resurrection — ISBN: 1-892427-04-4
- Volume II: The Early Church & the Middle Ages — ISBN: 1-892427-06-0

## All-American History by Celeste W. Rakes

Containing hundreds of images and dozens of maps, *All-American History* is a complete year's curriculum for students in grades 5 – 8 when combined with the S*tudent Activity Book* and *Teacher's Guid*e (yet adaptable for younger and older students).

There are 32 weekly lessons, and each lesson contains three sections examining the atmosphere in which the event occurred, the event itself, and the impact this event had on the future of America.

- Student Reader — ISBN: 1-892427-12-5
- Student Activity Book — ISBN: 1-892427-11-7
- Teacher's Guide — ISBN: 1-892427-10-9

## Christian Kids Explore Biology by Stephanie Redmond

One of Cathy Duffy's 100 Top Picks! Elementary biology that is both classical and hands-on. Conversational style and organized layout makes teaching a pleasure.

ISBN: 1-89242705-2

## Christian Kids Explore Earth & Space by Stephanie Redmond

Another exciting book in this award-winning series! Author Stephanie Redmond is back with more great lessons, activities, and ideas.

ISBN: 1-892427-19-2

**For ordering information, call 877-492-8081 or visit www.BrightIdeasPress.com.**

Bright
Ideas
PRESS